The Rational Scientist

October - November – December 2018

Copyright Grinning Monkey Publishing
All Rights reserved

ISBN-13: 978-1727303438

ISBN-10: 1727303431

THE RATIONAL SCIENTIST

A Quarterly Magazine

Available on therationalscientist.com in E format for immediate download, and on Amazon in 8.5 x 11 paper format.

Our mission is to introduce you to the Rational Scientific Method and to expose current mainstream nonsense such as Relativity, Quantum Mechanics, String Theory, Big Bang, Black Holes, Faster Than Light, Warped Space, Multi-Dimensions and TimeTravel.

Science magazines have gone astray. Today, the popular rags of science are more interested in scientific fantasy, and political debate or social justice than they are in explaining phenomena using objects. But, then, mainstream science abandoned using objects to mediate phenomena long ago in lieu of abstract theoretical mathematical descriptions and reification.

This issue deals exclusively with the extinction of ALL species.

"What Happened to the Dinosaurs" is a brief outline of the two main types of extinction; Mass extinction and Background extinction.

Economics is the management of resources and the currency of "Natural Economics" is food.

We covered climate change in the April issue of TRS. In "the Extinction of Man" we take a look at the claims of climate change alarmists such as Guy McPherson as it relates to man's extinction.

In "Political science: How Admixture Became the Law of the Land" Bill Gaede disposes of the Admixture theory for Neanderthal extinction.

In "Utopian Worlds - A Curious Experiment" Bill Gaede covers the mechanisms leading to the population pyramid overturning in his theory of Background Extinction. We learn that "Like all other animals, humans are subject to Mother Nature's laws."

We conclude this issue with a few chapters from the fiction book, "Bugworld." When a man matures he comes to understand that he will die. When mankind matures it will realize that mankind will die. This is not a call for wringing of hands but merely the acceptance of the inevitable: all species eventually go extinct. This is a glimpse of how it could go down.

Editor and Chief
Monk E. Mind

Executive Editor
Monk E. Mind

Creative Director
Monk E. Mind

Contributors
Bill Gaede
Monk E. Mind

Artwork
Monk E. Mind

COVER ART
Monk E. Mind

Published By
monkemind@gmail.com

CONTENTS

OCTOBER
NOVEMBER
DECEMBER
2018

03 WHAT HAPPENED TO THE DINOSAURS?

05 NATURAL ECONOMICS

15 THE EXTINCTION OF MAN

26 UTOPIAN WORLDS

21 ADMIXTURE

31 BUGWORLD

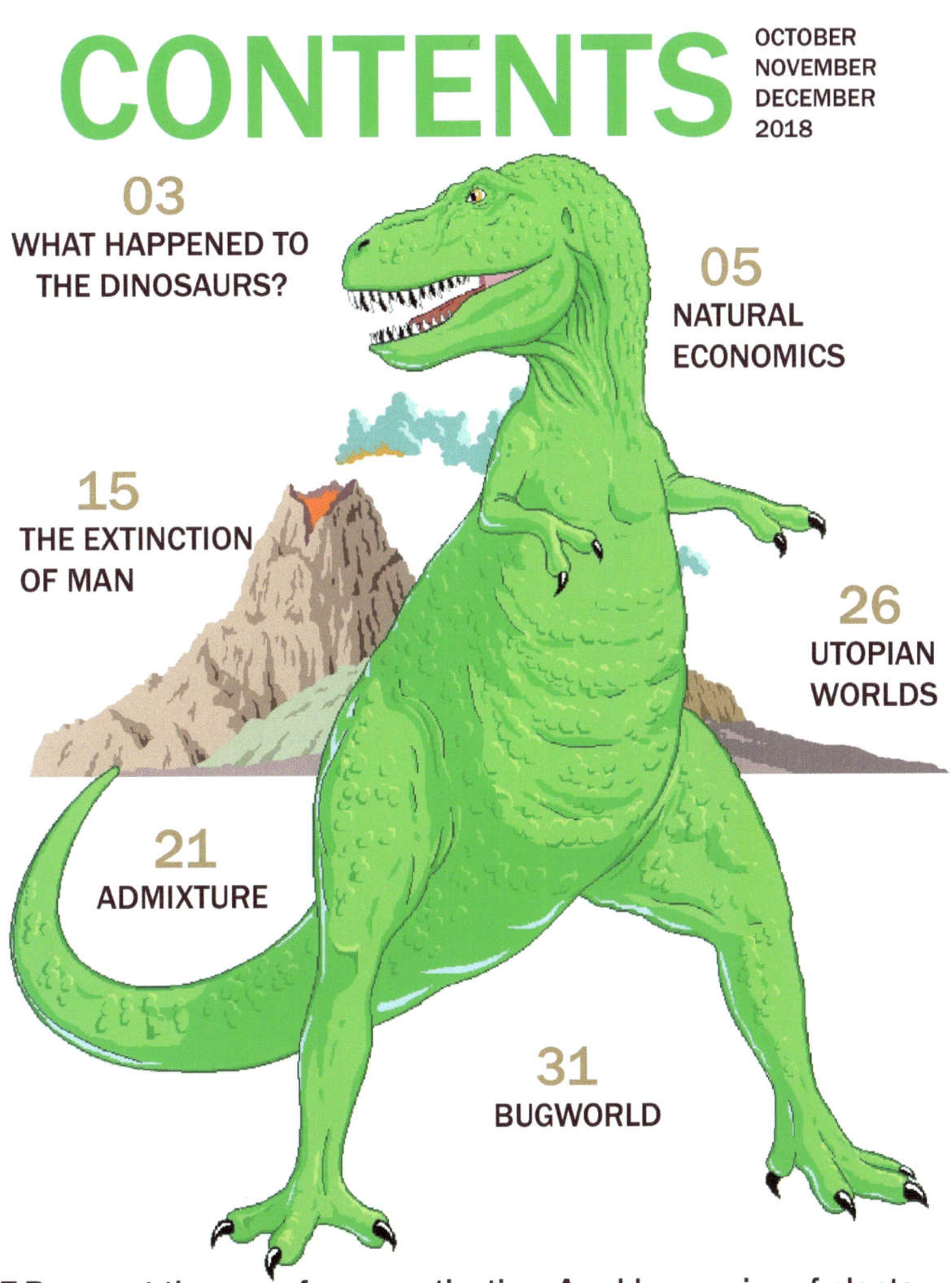

T-Rex went the way of mass extinction. As older species of plants died off, herbivores depending on those plants starved, and in turn, the carnivores that hunted them starved.

What Happened to the Dinosaurs?
By
Monk E. Mind

When I was nine years old I was kicked out of Vacation Bible School for "disrupting the class." The teacher told my mother that I was incessantly asking questions about dinosaurs during the reading of the Adam and Eve story.

After all these years, it has finally been made clear to me what probably happened to the dinosaurs. Today, I understand the mechanism behind it, the "laws of extinction." It reasonably explains why every species eventually goes extinct. It is either background extinction or mass extinction. No massive meteorite collisions with the corresponding impact winter. It wasn't disease, predation or any of the things that had been postulated and hypothesized before. Extinction is simply the overturning of the population pyramid or inversion of the ecological pyramid.

Amazing that no one had ever proposed the simple explanation of the natural economy until Bill Gaede did just a few years ago.

Economics is all about the management of resources, and for nature it boils down to the management of food. Unlike man's artificial economy, where, in this country, three percent of the population provides the food for the rest, other earth creatures have to fend for themselves.

Neanderthal reigned supreme. He was able to gather and hunt enough food to thrive. Eventually, the population grew old and died off by way of background extinction.

Background extinction is the overturning of the population pyramid, mostly as a result of density dependent birth rates and loss of genetic stock. The loss of biological diversity is the same for all plants and animals.

Declining diversity and increased specialization leads to a collapse of carrying capacity. The young were simply unable to provide for the growing number of old Neanderthals who couldn't carry their own weight. The population inversion resulted in the ultimate demise of Neanderthal. They simply died of old age. It's not the worst way to go, really!

T-Rex, on the other hand, went the way of mass extinction. As older species of plants died off, herbivores depending on those plants starved, and in turn the carnivores that hunted them starved.

During the Cretaceous period, Tyrannosaurus Rex relied mostly on Triceratops who ate mostly cycadeoids and cycads.

Tyrannosaurus Rex, the terrible lizard, was a carnivore. Triceratops, the three horned face, was an herbivore. Cycadeoids and cycads were types of seed plants with strong woody trunks and large stiff evergreen leaves.

Mosasaurs, large marine reptiles, ate mollusks, a type of shell fish such as clams. Mollusks ate plankton which are microscopic organisms like algae and protozoans.

During the Triassic Period, Prestosuchus with his serrated teeth munched on smaller animals like Hyperodapedon, a beaked reptile who scarfed mostly on Dicroidium, a fork-leafed seed fern.

As new plants developed, new species of animals developed along with them. With more plant diversity came greater animal diversity. As older species of plants reached their peak and died off, the herbivores that ate them died off.

The plants conditioned the environment for the next species of plants and animals that inherited the earth. This happened mostly by way of biochemical pumps that created small changes in the atmosphere over a long period of time, usually millions of years. As the atmosphere changed, new plants edged out the old plants, and the herbivores that depended on them starved, as did the carnivores that depended on them.

The nitrogen cycle and carbon dioxide cycles regulated the types of plants that proliferated. Cycads and cycadeoids were crowded out by other newer plants which thrived on the new conditions. Angiosperms were crowded out by gymnosperms. After Triceratops ate most of the cycads and temperatures dropped, new gases and plants arrived on the scene.

Cyanobacteria depended on a symbiotic relationship with cycads. The decline in cycads, led to a decline in cyanobacteria, which led to more decline in cycads. Triceratops populations declined and T-Rex starved.

This same process happened in a number of geologic periods, and this explains mass extinctions consistently.

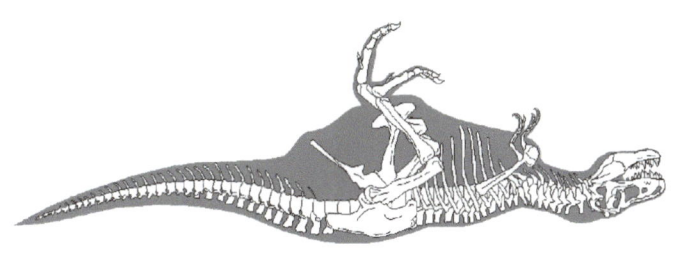

Natural Economics
By
Bill Gaede

I. Introduction

Recent discussions about the extinction of humans at our scientific forum (Rational Scientific Method) has revealed an unbridgeable gulf between two radically different ways of looking at Economics. On the one hand, we have what is typically known in our contemporary world as the discipline of Economics and which I will refer to as Artificial Economics. On the other, we find a long lost current, now exhumed, which I will call Natural Economics. The former is represented by the likes of Ludwig von Mises, John Maynard Keynes and Milton Friedman. The latter dates from the 19th Century and originated from the works of Lewis Henry Morgan, Karl Marx, and Friedrich Engels. **(1)** Their work has now been revised and is significantly different than what was initially proposed by these pioneers. Therefore, it is imperative to clarify that:

1. The term "Natural Economy" used by Marx and Engels also falls under "Artificial Economy" according to this article and in no way should be regarded as a synonym of the new concept of Natural Economy discussed here:

http://en.wikipedia.org/wiki/Natural_economy

2. The terms "Natural Economics" or "Natural Resource Economics" used by Artificial economists and their affiliates (e.g., the Venus Project) should also not be confused for the new concept of "Natural Economics" alluded to in this article:

http://en.wikipedia.org/wiki/Natural_resource_economics

Genuine Natural economists argue that the perspective originally proposed by Morgan and Engels and which now has suffered important modifications makes it obvious that:

we are undergoing the last stages of a mass extinction

2. The end result of this process is the imminent extinction of the top predator of the Earth: Man

This article is created as a reference to facilitate the discussions between these two irreconcilable currents of thought.

II. Density dependent birth rates (DDBR)

(2) Let us assume that Adam and Eve are placed in the Garden of Eden and they have unlimited quantities of food. They reproduce children mechanically every year and these children have children of their own every year. The population expands. God can wave His Magic Wand and make another miracle. He can, for instance, make it so the Earth increases its size, its diameter every time a child is born.

For any given cross section of time, the density of the planet is always the same. The amount of food – plants and animals they can eat – magically expands. There is always x number of people per km² and per pound of flesh and/or vegetables no matter what.

We could continue like this forever. Instead, let's be a bit more realistic and assume that the Earth and the quantity of food does not expand geometrically for every human born. Would Eve have one child per year forever?

Doesn't it seem logical that at some point the entire population of these predators will have to adjust to the availability food?

This intuitive or instinctive adjustment of the population to the amount of food available is known as density dependent birth rates. When we lived in the country and there was low density, women had children like if the Earth expanded. We had a long ways to go. Today we are close to 7 billion people on the planet and, although a great part of the world is empty, our populations are nucleated in cities, much like ants in ant holes. Since 1963, the global birth rate of humans has dipped from 2.4% to 1.2%. By 2050 it will be close to zero. Why?

Stress has been observed to be the factor that induces non-human females to forgo estrus. In the wild, this is triggered primarily by a dearth of food. If a female has a difficult time finding food, she has more important things on her mind than bringing another anchor into the world.

Like the newborn gazelle who never took a blackboard course inside her mother's womb yet "instinctively" "knows" that the cheetah has not come to celebrate her birthday, the tigress "instinctively" "knows" that she has trouble finding food in her territory.

The male tiger has in addition the stress of "instinctively" "knowing" that he will have to fight another gladiator for territory, either to expand or defend his own or to conquer his neighbor's. He can have fun with the girls and release the stress only after the war is over and only if the tigresses in the region have not crossed their legs.

If we define economy as "the management of resources," it is this vital resource – food – which underlies the mechanism of density dependent birth rates in animals. Mother Nature's economy consists of the management of food.

Our magical ability to plan and produce food, preserve it locally, and distribute it far and wide, has blind-sided the Artificial economist to the fact that humans are bound by the same laws of life.

Before a woman can decide to have a child she must be stress free. A 19th Century country woman waking up at the crow of a rooster had quite a bit of time on her hands and very few leisurely activities. If she and her husband agreed to expand their empire and ensure their old age, all they had to do was plant more acreage to cover the extra mouth. Having a baby was a no-brainer. It didn't require years of careful planning. Having large families was not due to the fact that our great grandmothers were ignorant brutes, which is what orthodoxy holds today. There was no reason for family planning because there were no cultural or economic obstacles to having clans. Quite the contrary. Culture practically demanded it and economics allowed it.

The labor that went into guaranteeing a new member of the family tribe was minimal and within the means of a farmer. It was not an undue drain on his resources to clear an extra plot of land, especially with some of the new machinery that was being invented, unless unforeseen drought or disease altered his plans.

Later, when a child grew, he or she would from an early age become another farmhand and contribute to the family economy which in turn made having yet another member again an easy choice.

The critical point, though, was that the couple was in control of their resources. They could decide how much to plant, and food was practically the only consideration.

I would guess that during drought or pestilence women probably held back. Aside from the cultural factors – religion, custom, peer pressure – having children was routine and not really a major decision. They could start having children as soon as they crossed puberty.

The new couple could live temporarily with their parents or be granted a plot of their own.

When the farmers migrated to the cities, all of this changed. I'll consider only the extreme, densely populated (locally speaking) Service Economy, just to make the contrast more visible.

A working woman is now limited by resources she has no control over. It is her boss who can hire or fire her, determines her salary, and can offer her overtime and bonuses. Her social security – the most important consideration for marriage and child bearing decisions in our crowded ant holes – is in the hands of others.

Unlike the calm country girl, the urban woman is under constant stress, 24 hours being insufficient to do it all. The meager wages she earns must be divided up to pay for the "necessities" of city life – from basics such as water, electricity, and rent to transportation, clothing, education, leisure, socializing, and food. Her paper thin disposable income is what remains to invest in child – as in singular, which although grammatically incorrect is infinitely more realistic than the plural "children." And, as just mentioned, she has practically no control over her paycheck! Her income is her ration, the amount that has been allocated to her by the Service Economy. Our Artificial Economy can't be more generous for fear of subtracting from

Do you suspect that something is wrong with what you read about in the so-called physics journals and popularization magazines? Don't black holes sound a bit magical? Doesn't Big Bang come across as Creationist? Does the establishment's "explanation" that there are many copies of you, each in a different universe, make you wonder about the state of mind of the theorists? If so, perhaps it is time for you to consider an alternative.

Why God Doesn't Exist (WGDE) presents a fresh perspective on the nature of light, gravity, magnetism, the atom, and the workings of the Universe in general that is rational. It is rational because each theory is illustrated. You can understand the mechanism by just watching the videos that accompany the book. It is also mainstream theories, you will not be asked to believe in the movement of concepts (e.g., transfer energy, move "a" mass, carry "an" interaction). WGDE is for intelligent laymen who simply want to understand the causes and mechanisms that underlie physical phenomena.

To obtain a paperback, Paypal to bill@youstupidrelativist.com

USA/Canada: US$ 30   Europe: 30 Euros

efficiency. An urbanite lives at the cutting edge of survival. For all practical purposes she is a slave. She has enough to live one more day so that she can make it to the rock pile tomorrow.

Aside from the inevitable cultural and spatial factors imposed by the densely packed Service regime – bachelors, single parent, homosexualism, divorce, density, apartment, finding the blue prince, unfaithfulness, insecurity – these are the real, day-to-day economic variables that distinguish the large family country mother from the lonely city wife. These are the realities the urbanite faces daily and which affect her childbirth decisions.

Bacteria growing along the S-curve on a piece of 2-week old piece of bread are more or less in the same boat, and if they are, it is hard to believe that animals in the evolutionary hierarchy above them are not. This observation presents us with an outline that explains density dependent birth rates for any living entity, especially, for large beasts. Although the T-Rex and the Neanderthals never even remotely attained the numbers we have, they were also subject to density dependence at their own levels. It has been settled that human hunter/gatherers had very low densities.

There absolutely must be a ballpark 10 to 1 ratio between trophic levels for the ecological pyramid of the Wild Kingdom to stay healthy and sustainable. If the top predator agreed to adhere to this ratio, so did everyone else in the chain of command. Therefore, DDBR is the key connection between population and economics. If we define extinction as the complete elimination of the population of a given species, Economics seems to be a discipline that should be considered when analyzing extinction.

More to the point, if as has been established, a mass extinction is patently the disappearance of an entire food chain – an entire ecological pyramid – then Economics is a sine qua non factor in these types of extinctions.

These conclusions demand that both biologists AND economists be summoned to the next extinction conference.

III. Artificial Economics

A mathematical "physicist" believes not only that marbles roll down the trampoline of warped space, but holds that if the Sun were to disappear it would take 8 minutes for the Earth to say goodbye to its orbit.

He also entertains the possibility of traveling through time and wormholes, searches for magical black holes through his telescope and for particles of mass with his accelerator, and has concocted the most preposterous principles ever to come out of the mind of Man – Anthropic, Uncertainty, Complementarity – to explain physical phenomena which to him still remain enigmatic.

Similarly, an adherent of either the Chicago and Austrian Schools of Economics has the eye-popping belief that increasing investment magically creates jobs. Absent from such sweeping declarations are so much as a hint of qualitative or contextual considerations. Friedman, Keynes, and von Mises types were born and raised in Disneyland and naturally take for granted and have always assumed that Mickey Mouse and Donald Duck are for real. They stand in awe when a skeptic even mentions in passing the possibility of the contrary. The artificial economist can't see past the perimeter of Wonderland. He has been dazzled by lights that he can magically turn on by flipping a switch on the wall and by smartphones that instantly connect him with people on the other side of the planet. The walls of his apartment are the extent of his intellectual horizon.

Since he has done everything through the magic of money since birth, a Disneyland economist knows no other world and solves all problems with this supernatural medium, firmly believing that it grows on trees. Asked by an engineer how he would solve the finite amount of oil available on the planet, the artificial economist is likely to answer that it's just a matter of price. Milton Friedman synthesizes this worldly wisdom in his dictum: "Most economic fallacies derive from the tendency to assume that there is a fixed pie..." In Friedmanian Economics, scarce resources are physically made to appear in infinite quantities by throwing money at the pies. Thus, it comes as no surprise that a Friedmanian casually believes that investment automatically equates to jobs. There's no attempt to factor qualitative or contextual parameters into the equation.

Therefore, if God were to wave His Magic Wand today and wipe out all the money on Earth, a rational person would answer that all the thin threads connecting the global web of relations would be irrevocably severed and the spider would fall. There is no way to stitch them back together. The Artificial economist offers, instead, a stunning answer. He tells you that he would replace the old green money God took away with new brown money or with another medium of exchange, including bartering. That's when it suddenly dawns on the startled inquisitor that he was not talking to the psychiatrist at the asylum, but to the patient. Such breathtaking replies irrevocably separate the handful of down-to-earth Natural economists from the irrational hordes who follow the likes of von Mises, Keynes, and Friedman. **(3)** There can be no reconciliation between these two currents of thought.

In retrospect, as happened in Mathematical "physics," it was predictable that a surrealistic language would evolve that would enable the Artificial economist to manipulate and juggle concepts in his lab. The Friedmanians don't talk about tangible sticks and stones, but about supply, demand, savings, investment, taxes, interest, debt, labor, money, and other mass nouns that exist in Wonderland. It's a dialect of abstractions that enables the contemporary Artificial economist to grasp and talk about the Service world in which he lives. All members of this religion speak the same language, yet inexplicably accuse each other of heresy. The von Mises Sect sneers at the Friedman Sect which laughs at the Keynesian Sect... and all of them at the Marxist Sect.

IV. The supply and demand curve for food

A supply and demand curve pits price against quantity. Typically, the more of one, the less of the other. To an adherent of the Friedmanian school of thought, every tangible object, any lump of atoms, can be bartered and sold. It's just a matter of price. And it is patently obvious to him that a Cadillac is worth more than a carrot.

No contest. He has this opinion, however, because I have not yet described the context in which I ask him to choose. The context is that he is marooned on a deserted island. Which would he choose now? Which would he want to have after three days? A carrot or a Cadillac? This example underscores the unbridgeable chasm that exists between the members of Disneyland Economics and those handful of rational, Natural economists that still remain on our planet.

(4) Here's an example more closer to home. What would happen to you if you did not have any money? What if you lost your job, your house, your wife and dog and became a bum? Would you turn in the bottles that you picked up in trashcans to feed the one-armed bandits at the casino or to buy food? What do you think would be important to you?

(5) From Mother Nature's perspective, food is unlike any other commodity manufactured by Man. Firstly, it grows; it is not assembled. Then, it is deliberately grown in proportion to population. This enables us to replace the traditional price axis with population and the quantity axis with weight or poundage. As long as population remains constant, the supply and demand lines for the entire planet will barely slide.

In other words, food is a Goldilocks type of commodity. You can't have too much or too little of it. It has to be just right... under any type of economic arrangement, whether Hunter/Gatherer, Agriculture, Manufacturing, Service, or Unemployment. Not so with other commodities. We can reduce the number of radios we demand and have reduced Pet Rocks to zero. These crucial distinctions between grown (necessary) and manufactured (unnecessary) tangibles become relevant in scenarios such as the one depicted above. If your ship capsizes and you swim to the nearest island, a car or a TV set is probably not be the first thing you want to take with you ashore.

V. And the problem is...?

Why would a Natural economist even raise such concerns? "Where is all this coming from?" asks the economist of Wonderland? Has a boat capsized? Are we marooned?

They are important because our Artificial Economy, the one we experience every day and which has blind-sided the Disneyland economists to the above facts, produces food not in relation to the number of mouths, but in response to market conditions. The two are the same in the Artificial Economy. They are not the

The Best Of
Rational Science

Monk E. Mind

same in Mother Nature's Economy. Like our intellectually disadvantaged friends said, the Artificial Economy treats food just like any other commodity. Vertically integrated agricultural multinationals such as Archer Daniels Midland, Nestle, and Bunge process and distribute most and increasingly more of the world's food. **(6)** These companies are not in the business of feeding you. They are in business to make profits. They are impersonal companies doing business just like any other, selling a tangible commodity for a price. If as hypothesized earlier, God suddenly eliminates all the money in the world with His Magic Wand, these companies have no further purpose for being. They have no incentive to grow, process or distribute food to the world. We can do without TV sets and perhaps even computers for a few days and if necessary forever. We cannot do without food for more than a few days. You wake up one morning, go to the supermarket, and there is no food. The Artificial Economy is no more, and you come face to face with what every animal in the Wild Kingdom experiences daily: Mother Nature's Economy. The only relevant resource that animals manage is food. Without it, they die. Without food, humans die in the midst of Cadillacs, mansions, highways, skyscrapers, computers and all the "unnecessaries" that the Artificial Economy built. The only thing we need is food. The Disneyland and Wonderland Worlds are still there in front of us for everyone to touch and see, but we suddenly discover that Mickey and Donald and the Queen of Hearts had all along not been real. We discover that we lived in a fake, artificial world that now cannot help the reality of hunger we feel in our stomachs.

Why would money disappear overnight?

It would disappear because the Man's interrelated, GLOBAL Artificial Economy is that close to collapsing. And here there are irreconcilable differences between Artificial and Natural economists as well. The Artificial economist has been raised and educated to believe that everything is a cycle. He gets his pointers from his brother, the mathematician, who holds philosophically that ultimately everything is relative, meaning, a matter of opinion. This explains in great measure why Einstein's amusing "warped space" theory was widely accepted. Some relativistic mathematicians believe that the contraction of a ruler that travels near the speed of light is real. Others, that it's merely a mathematical expression of the frame of reference. Some believe in black holes and Big Bang. Einstein himself vouched for neither. The beauty of relativity is that everyone is correct.

In like manner, since the Wonderland economist never factors the 100,000 year history of Man, but at best the last 100 years of history, he has come to the conclusion that Artificial Economics is but an endless cycle of business booms and busts. He doesn't realize that the context in which all this history happened comprises only a very short strip of history: the last stages of Manufacturing and the whole cycle of the Service Economy. The Artificial economist never ponders the long term history of Man. When the entire history of the human economy is analyzed, there is clearly a linear pattern. Man went from Hunter/Gatherer to Agriculture to Manufacturing to Services and never went back through any of those modes or categories.

(7) Thus, when the Artificial economist insinuates that the global economy will or may collapse, he is referring to another of his cycles. He has in mind a Depression Era style period of lean cows and thin wheat stalks from which we will eventually recover, perhaps under some unspecified new socioeconomic relations. This is in fact what the followers of Ludwig von Mises (Austrians) and Jacque Fresco (Venus/Zeitgeist) prescribe. When the Natural economist talks about the global economic collapse, he is referring to a dead end scenario, a complete and irrecoverable collapse of money. The threads of the global web can no longer be mended with any artificial waving of the Magic Wand. Issuing money of any kind will not reestablish the broken links. The global economic collapse results in the disappearance of all governments as well as of the entire chain of command, military included.

VI. I still don't see the problem

So what is the problem?

The problem is that not one Artificial economist can imagine what job category comes after Services. If as the long term, linear, irreversible trend went through ever more efficient stages, what is the next phase? Where will billions of workers work when the Service Economy, which is becoming more and more efficient as I write, sheds half of its labor in pursuit of cost cutting and innovation?

After a moment's thought, the Artificial economist realizes that there is no category he can think of. Realizing that there is a long term trend and being unable to think of the next category, he changes tactics. He asks, "Why do we need another category? or tells you flatly that, "We don't need another category." Again the Disneyland economist reverts back to the last 100-year cycle: "What the Paleolithic hunter/gatherers did 35,000 years ago or the Neolithic farmers did 5,000 years ago has no bearing on the economics of today."

Such replies indicate to the Natural economist that he again mistook the patient for the shrink. The economist of Wonderland simply cannot see beyond the walls of his apartment. That's the extent of his intellectual horizon. He will revert to his endless loony asylum ranting where he prescribes raising investment to create jobs out of the tip of his magic wand or lowering the tax rate to create incentives for businesses to have more disposable capital for research or vehemently argue that we need to get government off our backs. There is no consideration in his entire babbling for qualitative factors. He has not realized that, on the one hand, competition in an Artificial Economy NECESSARILY conduces to layoffs, and on the other, that there are no new subcategories of Services in which to shift these countless laid off workers. Yet he asks, "Why do we need a new major category?" After a minute of thinking, he improvises next that we will invent something, typically, robots. This, he believes, will create jobs for millions of blue collar mechanics and technicians. He has again solved the problem with his magic wand. He is proposing that humans will go back to Manufacturing. He has forgotten that Manufacturing has been leaking workers into Services now since World War II. There were only 12% of workers working in the Service Sector in 1941. Today, it is over 80%. For the world, the labor employed in the Service Sector is estimated to be around 45%. And growing! Ergo: the Service Economy.

Other Artificial economists propose the contrary: going forward. The future society is a utopian world in which remote controlled machines do everything. We will evolve into the Leisure Economy. Again: magic. You wonder why all these people ever went to college.

The global population birth rate has been steadily declining since 1963. It stood then at 2.4% and today stands at 1.2%. UN demographers have projected that by mid 21st Century the global population birth rate will grind down to 0, a phenomenon that is known as Zero Population Growth (ZPG). **(8)** This estimate has been revised downwards year after year. It originally was supposed to happen in the early 22nd Century. The Artificial economist casually proposes that the population will oscillate around 8 or 9 billion people for the rest of eternity. The only two other options are increase without limits and decrease until extinction. **(9)** Which of these three trends will Man follow?

ZPG is a consequence of DDBR. Humans have nucleated in cities, and economic and cultural factors inherent in urbanization prevent the average woman from having many if any babies. When we consider these two symbiotic factors – DDBR and the childless bachelor (the typical "family" unit of the Service Economy) – we realize that unless the bachelor unit reverts back to the nuclear family, the population has no chance of recovering. However, the nuclear family is the unit of the Manufacturing Economy, an era long gone. No invention will ever again put billions of people to work like the automobile, telephone, light bulb and other such ground breaking inventions did in the late 19th and early 20th Centuries. Part of the reason is that in the incredibly efficient Service Economy – the Wonderland in which we are born – people have or quickly obtain practically anything they need or ever wanted in order to live comfortable lives. Therefore, those who propose that humans will oscillate back and forth around 8 or 9 billion people forevermore have to answer how ever increasing urbanization is going to result in an increase in birth rates in the context of the Service Economy's bachelor society and DDBR.

If, in addition, we now factor that the Service Economy cannot be run on a constant population, that businesses absolutely need to sell ever more goods and services rather than just increase prices to make a profit, we realize that the global economy WILL collapse at some point. The Artificial Economy is an inverted pyramid scheme. It is sustainable only if there is ever more demand (i.e., population). The day population comes to a halt, so does the Artificial Economy.

VII. Conclusions

The unemployment we see and hear about in the news today is not cyclical as the shortsighted Artificial "economists" will have you believe, but structural. The reason for the unemployment you read about today is that Services is shedding employees to produce profits for its investors. These ex-employees go first to the unemployment lines and then to welfare. This trend is unsustainable.

The Artificial economists, those Friedman types who solve every problem with money and magic wands, will tell you that they will create jobs by increasing investment and that we will just replace the green money God took away with brown money and continue as if nothing had happened. The Natural economists will argue that all the threads of the global spider web are now beyond repair. Creating new money will not mend the connections. If it were that easy to create jobs by merely increasing investment, the governments should be able to solve our current problems easily. All that the politicians need to do is create money at will from the end of their magic wands and put people to work today. Indeed, if it were that easy, the governments might as well create money and give a million dollars and euros and yen to each individual. Then, none of them would even need to go to work.

Footnotes

(1) Although the work of these authors dates from the late 19th Century, many of their concepts, perspectives and findings are still relevant in the opinion of the Mother Nature Economists. It is recommended that those wishing to enter the debates familiarize themselves with the underlying arguments at:

http://www.marxists.org/archive/marx/works/1884/origin-family/

(2) I am indebted to my niece Mercedes Gaede for some of the key points in this section after our productive exchange of ideas on the morning of Jan 8, 2012.

(3) I lump the followers of von Mises, Keynes, Friedman and all the other modern economists in the same basket and refer to them alternatively as the Friedmanians, Artificial economists, or Disneyland or Wonderland economists because there are no differences between these schools of thought from the anthropological perspective of Morgan and Engels.

(4) A Natural economist has no use for any form or shape or concept of trade or bartering. There is no exchange of goods or money in a Natural Economy. Mother Nature's Economy consists solely of food. In the wild, no animal exchanges food with any other animal. In this, the contemporary Natural economists distance themselves not only from the Artificial economists, but from the likes of Marx and Engels and the analyses and solutions they prescribed in works such as Das Kapital as well.

(5) I owe this scenario to my youngest son Adrian who proposed it as a means to illustrate to the new generation what the collapse of the economy entails.

(6)
http://en.wikipedia.org/wiki/Industrial_agriculture#Historical_development_and_future_prospects

(7)
http://youstupidrelativist.com/08Ext/09Econ/04EconHist.html

(8)
http://www.census.gov/population/international/data/idb/worldgrgraph.php

(9)
http://youstupidrelativist.com/08Ext/05Demog/01PopTrend.html

Rope/Atom Lattice T-Shirt
Annoy Your Professors
Delght Your Friends And
Dazzel the Ladies
100% Organic
Men's Fitted T-Shirt
Get Yours Now!
cafepress.com/rsoutley.1798637511

The Extinction of Man

On June 25, 2017, a follower of Guy McPherson, Jackson Davis, posted several links to Guy's material including an Interview entitled, "Near-Term Human Extinction #218" to Bill Gaede's facebook group, "The Extinction of Man." The following is some of the ensuing discussion slightly edited. Find the original conversation here:
https://www.facebook.com/TheExtinctionOfMan/posts/2325622027662852

Bill Gaede People never understand the theory that is put before their very eyes because they don't read. Instead, they ASSUME they know what the instant theory is about and post nonsense such as that humans will become extinct because of climate change, environmental havoc, overpopulation, diseases, peak oil, or other NONSENSE that has NOTHING to do with extinction.

See *"We are the last generation of humans on Earth, Proc. Int. Oxford" Academia.edu*

Jackson Davis If you read this article by Dr. Guy McPherson, you will see that the collapse of civilization is a part of his prediction that human extinction will occur within 10-34 months from now:

Bill Gaede In Science, we don't do predictions. And the instant THEORY is NOT a prediction. The End of Man WILL happen in the future, but it is NOT a prediction. And of course, to the person who was brainwashed with Mathemagix, this seems like a contradiction because he is unfamiliar with the Scientific Method. See *"What is Physics? Science 341" (2014) Academia.edu*.

Luis A Gonzalez There is not going to be any extinction of man for a long time.

Bill Gaede That's just a subjective OPINION. All you did is kick the can down the road: "It will not be our generation. It will be some future generation."

So now... for the juicy part: the theory. Present a full and fool proof mechanism for the extinction of Man. What agent or process WILL INEVITABLY kill a future generation, Luis?

Guy McPherson *Edge of Extinction: I'm an Optimist? Nature Bats Last, guymcpherson.com*

Luis A Gonzalez Foreboding, the give away is the time prediction. 2025 will come and the planet and humans will be fine.

Bill Gaede Dear, dear Luis. 2025 is as good year as any. It could even happen sooner. You cannot justify your optimism. And again, Science is not about making predictions. That's the hallmark of the religion of Mathemagix. In Science, we CONCEPTUALIZE what will happen to Man (no matter what) just like a grand master conceptualizes how the chess game will end irrespective of how he moves his pieces in the next 5 moves.

The problem with GM's Edge of Extinction is that he doesn't have ONE theory. He brainstorms all imaginable causes to cover every base. So he throws in there how the rich abuse the poor, climate change, nukes, peak oil, disease, US military killing less fortunate folk, and whatever he can sift from news clips and mainstream extinction buffs. Therefore, he comes across as Chicken Little screaming that the sky is falling. He's not a realist, but rather a true blue pessimist. So let's put the cards on the table. Climate change and peak oil have NOTHING to do with extinction. They should be removed from the list of causes. No one should ever mention such retarded causes ever again. Who cares if the polar ice caps are melting or if we pollute the environment or if global temperatures rise 5%, if, as he suggests, we will all die by 2025? We will never get to experience any effects from global warming in what little time we have left.

And who cares if the panda becomes extinct or that we kill all the lions and gorillas? Who needs them? In what way will their disappearance and that of so many others affect us? Did the extinction of the mammoths perchance trigger the demise of Man?

And I have a hard time believing that the dinos died because they ran out of oil or because their vehicles emitted too many carbos. Nor am I convinced that the T-Rexes threw nukes at each other. They died for one reason and one reason only. Their sources of food disappeared. G.M. mentions starvation in passing, a mechanism he pulls out of nowhere. He doesn't justify how or why starvation will happen. He also mentions "industrial civilization reaching an end," again, without justification or cause. Why will it? Do we run out of oil? Did he get that from Jay Hanson's site Dieoff?

He should first learn that "industrial civilization" passed away 40 years ago. Practically all countries -- certainly all relevant ones -- have morphed into Service Economies and they are now phasing into the Unemployment Economy. As a minimum he should modify his language.

Starvation will happen and will be the cause simply because we can't have ever more unemployed (especially the young folk) being maintained by ever fewer employed. At some point growth stops and the global econ collapses. The golden motto of economics is: "Grow or Die!" The rich and famous, the Illuminati, the powers that be, the Skull & Bones, the New World Orderers, The Masons, etc., none of them can do anything about it. The global economy WILL collapse at some point and there is no antidote. This is not a prediction. This is a conceptual issue. The date this will happen is IRRELEVANT to CONCEPTUALIZE that we are going to crash against a brick wall. You get distracted by the date and the "prediction" and miss the point: we WILL become extinct. It will happen SOON. I hope that's an accurate enough prediction for you.

Please do not confuse me with Guy McPherson. He is a conservationist, a climate-changer, an individual who has no clue whatsoever about EXTINCTION... It is IDIOTS who believe that humans will go extinct because of the environment or pollution or because we are too many.

The dinos didn't disappear because the climate changed or because an asteroid struck the Earth or because of pollution. They went extinct like humans will go extinct: STARVATION! It is when food runs out that a mass extinction occurs.

Jackson Davis This is Dr. Guy McPherson's latest:

Abrupt Climate Change Leading To Near Term Human Extinction Here is the outlook of Dr. Guy McPherson, Professor Emeritus of Natural Resources and Ecology and Evolutionary Biology at the University of Arizona, and the world's leading authority on Abrupt Anthropogenic Climate Change leading to Near Term Human Extinction:

"We are now at a Global Average Temperature of 1.73 C above the 1750 baseline and it is apparent that it has started to increase exponentially. An 'ice-free' Arctic Ocean later this year or the next increases the probability we'll have a sudden release or "Burst" into the atmosphere of 50 Gigatons of Methane from the frozen Methane Clathrates in the Permafrost lying on the shallow seabeds of the Arctic Ocean causing the Global Average Temperature to rise 1.3 C more in a very short time.

"This actually means a more than 2 C Temperature Increase in the so-called 'Breadbasket Regions' of the Northern Hemisphere (North America, Europe, Ukraine, and Russia), where Most of the Plants that Humans and Other Animals depend on for Food are grown. The effect of this Temperature Increase is that these Plants will Die and Mass Starvation will be the result.

"Mass Starvation in turn will mean a Collapse of Industrial Civilization and a steep Reduction in the Burning of Fossil Fuels. This might seem beneficial at first, since there would be a Reduction in the Emissions of CO2 and other so-called 'Greenhouse Gasses,' but the Global Average Temperature will continue to rise due to the Emissions Generated over the last 10-20 Years.

"However, the Reduction in the Burning of Fossil Fuels will also result in the Reduction of the Emissions of Aerosol Particulates such as SO3 (particularly from 'Dirty Coal') which REFLECT Solar Energy back into Space before it reaches the Earth (the so-called Global Dimming effect). These Aerosol Particulates Fall Out of the Atmosphere in a matter of Weeks or Months and the Global Average Temperature will rise another 3 C causing the Extinction of Virtually All Living Beings on Earth. This could happen by next Spring or the Spring of the following year, so be prepared.

"At The Edge Of Extinction, Only Love Remains!"

Bill Gaede Climate change never exterminated a single species in the history of life on Earth! A mass extinction is ALWAYS, in every instance, a case where primary production dwindles to unsustainable levels. And before the appearance of modern Man, all background extinctions were the result of the overturning of the population pyramid of each species. Climate has nothing to do with either mechanism. He who invokes climate change in the context of extinction is nothing but a misguided Chicken Little.

Jackson Davis Our "pyramid" is about to be overturned !

Bill Gaede YES! But it has NOTHING to do with how humans are going to go extinct!

Jackson Davis "Primary production declining to unsustainable levels" is precisely the result of Abrupt Climate Change and that in turn will result in Near Term Human Extinction.

Bill Gaede NO! Absolutely NOT! Climate change has nothing at all to do with it. We humans produce the food we consume. We do it for profits. We could even produce more food if it was "worth" it... despite any climate or other environmental change.

It is MONEY which produces food. Without money we die tomorrow morning. Our lives depend on this abstraction that humans invented to keep themselves fed.

The artificial economy that humans have created is about to reach a dead end. There is no way the global economy can continue growing eternally. At some point unemployment and debt will be so high that the lack of global GDP growth won't be able to cover the hole by just printing money. In God we -- no longer -- trust. The greenback is laid to rest. Money will have come to an end. And without money there are no profits. And without profits there is no incentive to produce or distribute food. The cities starve to death. 8 billion people die in a blink of a geological eye... in a matter of weeks... just like the last dinos... It's all about money and none whatsoever about climate. The dinos died because their ECONOMY collapsed. There was no longer any resource to manage. The only resource of any importance that an animal manages is food.

Just in case -- because some don't have the intelligence to figure it out -- the first to die will be the lost tribes living in the jungles. The local city folk will rush to where there is food. They will hunt down whatever the natives hunted and then hunt down the natives themselves.

Jackson Davis You see, even if there were no Methane threat, a collapse of civilization will result in a sudden reduction in the burning of fossil fuels. Yes, this would result in a corresponding reduction in emissions of CO2 and other greenhouse gasses, but the temperature would continue to rise because of all the greenhouse gasses that have been emitted during the last 10-20 years.

But there also would be a sudden reduction in the emission of aerosol particulates, such as SO3 from burning "dirty" coal, which REFLECT solar energy back into space (the so-called Global Dimming effect). these particulate fall out of the atmosphere in a matter of days or weeks and the temperature will quickly rise 3 C. This is enough to bring about the extinction of humans and almost all higher life forms in a short period of time.

Bill Gaede Some people believe that EXTERNAL, EXTRINSIC factors such as CO2 or global warming or runaway greenhouse or pollution or the weather or asteroids or volcanoes cause extinctions. That's what they learned to memorize by rote. That's what the "experts" told them for years. And that's what Guy McPherson also memorized and repeats.

ALL background extinctions before "civilized" Man without exception were caused by the Overturning of the Population Pyramid. The species died of OLD AGE!

ALL mass extinctions without exception are caused by the Overturning of the Ecological Pyramid. The PLANTS that the reigning species eat undergo THEIR Population Pyramid Overturn and disappear. The dinos vanished when the cycads gave way to the angiosperms. The Permian Mass Extinction occurred when the ferns gave way to the conifers. The Carboniferous amphibians disappeared when the club mosses disappeared. And so on. ALL mass extinctions are due to FOOD!

But folks like Jackson Davis and Brian Edwards and Guy McPherson are gawking at the sky... to see if the air is clean... or at the ice... to see if it melts... or at the number of people there are... to see if we exceeded the carrying capacity of the planet!

Man will go extinct soon and for reasons that have nothing to do with climate or CO2 or volcanoes or asteroids or any of the other NONSENSE that the "experts" spread out there. Why... there are even people out there who believe that humans mated with the Neanderthals because we have 2% nDNA in common. Can you believe such garbage? And such nonsense comes from "experts" such as Svante Paabo. And people believe such nonsense only on the basis of authority.

https://www.youtube.com/watch?v=CsH_QGqFWBA

Man dies WHEN his economy finally disappears and MONEY is no more. We need not worry about the air we breathe or the melting of the ice caps or the disappearance of species like the Woolly Mammoth, the Passenger Pigeon, the Great Auk, or the Thylacine. In fact, we should wipe out the lions and sharks and bears and crocs and all other dangerous animals today! We don't need them. We won't miss them. The world can continue without them.

So what do conservationists do? They do the reverse. The environmentalists attempt to save the panda and the tiger and the cheetah... when humans can't even save themselves!

See Neanderthals: Against Admixture 03 mtDNA

https://www.youtube.com/watch?v=CsH_QGqFWBAfs

Jackson Davis I suppose the dinosaurs also became extinct because they ran out of money.

Bill Gaede The dinos ran out of food. That's the economy they managed. Humans also die when they run out of food. That's the only resource they ultimately manage. ALL the other things humans produce are UNNECESSARY and only blind us to this fact. All we need is food to stay alive. We need no computers, cars, or services to stay alive. It's the biological and physiological stuff that kills us. We just need food, air, and water... just like any other animal. Instead, we think that we're special, that God designated us as the caretaker of the Garden.

Humans manage their economy through money. It is money that grows wheat and corn in the fields. Without money there wouldn't be crops or meat -- we're way past subsistence agriculture.

The global money economy comes to an end when growth finally stops. Unemployment and debt will have reached their respective ceilings. Creating more money or brown money or yellow money at that point will have no effect. It won't create jobs. And it won't pay the creditors in real value. Money will be what it is: an abstraction, a bunch of numbers in a computer. This is the only fragile "thing" keeping the world alive at this moment. People don't realize this.

And the global economy is THAT close to total collapse... It can happen at any moment. And yet people believe that the melting of the ice caps will kill Man... or perhaps a disease... or maybe an asteroid...

Jackson Davis If you believe that as long as you have money you can survive, try counting your money while holding your breath.

Bill Gaede Won't put food on the table...

Gaurav Jaiswal Climate change can greatly affect agriculture. When you can't grow crops, it's over. Food will become expensive and then there will be riots and chaos. Only impossibly rich people will survive, slavery will return and life will continue for humans. All progress will come to halt. It will be a long time before we finally go extinct though.

Jackson Davis Do you have evidence or is that just wishful thinking?

Gaurav Jaiswal It's a plausible scenario. As long as people are able to feed themselves they will compromise in whatever living conditions they are in. So unemployment and economic breakdown is nothing to worry about. But when the crops won't grow and you have to grow them in a highly regulated environment, it will require a lot more energy and a lot more work. Food will become so expensive that even a middle-class income would be insufficient.

THE RATIONAL SCIENTIST

JANUARY FEBRUARY MARCH 2018

What is the Rational Scientific Method?
page 3

WHAT IS RATIONAL AND WHAT IS LOGICAL
page 6

RATIONAL PHYSICS
page 11

TECHNOLOGY
page 12

Bill Gaede INTERVIEW
page 19

Bill Gaede "unemployment and economic breakdown is nothing to worry about"

Famous last words...

If the Almighty waved His magic wand at this moment and made money disappear (rather than drown mankind with rain), who would produce and distribute food?

People have really taken food for granted! And this is because food appears by way of the same magic wand on the supermarket shelves. It's always "there." Who knows how it got there, right? Magic. You just have to take money and it's yours, right?

"When the crops won't grow and you have to grow them in a highly regulated environment, it will require a lot more energy and a lot more work."

Been there, done that! For money, we can even terraform all of Mars! Humans can deal with anything Mother Nature throws at us: floods, drought, tornadoes, tsunamis, ice cap melting... Where Mother Nature won't intervene is in Man's economy. She doesn't understand numbers. She never went to school. She can't help us there.

Jackson Davis "Where Mother Nature won't intervene is in Man's economy." Now that's REALLY funny !

Bill Gaede Exactly! That's totally man-made! We went from hunter-gatherers (200,000 years or so) to agriculture (10,000 years) to the manufacturing economy (300 years) to the service economy (50 years) to the current unemployment economy. And that certainly can't last more than 50 years. Mother Nature can't put us to work. She can't create economic growth. She can't take us back to agri or to hunter-gathering. And she surely can't create more money forever and keep us alive so artificially. Quite the contrary! She only allows a species to be king of the planet for a day. Our time is up! It's time for the roaches or some other insect to rule the planet until it is gone as well.

Jackson Davis My point is that Mother Nature can devastate the economy.

Bill Gaede No. She can devastate the environment and make it hard on us. We can overcome whatever she throws at us... as long as money is alive...

Gaurav Jaiswal People will create another currency. Easy! Who produces and distributes food? Humans. It's a chainlink. Farmers grow food, they sell it in the market, market sells it to the companies that process it and then there is another market where you go to buy the processed food. The problem is that people will never stop using money. There is no way money will lose it's value because everyone in this chainlink benefits from money. Why would people lose faith in money, when they know that without it they won't be able to buy things and food and that there would be complete survival of the strongest without it. Strong people will form clans and start doing their own production and grow muscle and thrive. Enslave weak and have them work in the farms and factories. Still, no extinction, unless World War Three in the mix.

All technological progress would come to halt and law of the jungle will be put in place. A great depopulation event would occur but no not the extinction. Humanity have survived without money and even before there was any money.

As long as we have the means to produce food we will survive, but a drastic climate change can take that away and make it harder for us to do that. Now without money you have these clans that own hoards of slaves but control different sectors and occupy resources that they exchange in order to continue the new way of life. End of money is not end of humanity.

Bill Gaede "People will create another currency. Easy! ... people will never stop using money."

Ha, ha, ha. You certainly are an expert in economics, GJ. I stand in awe. In fact, thanks to your post I now have a way to save the world. I plan to write to every Gov on Earth and ask the powers that be to give each citizen 1 million dollars! That should make every person on Earth quite happy and, as a by-product, solve all of the problems the Govs have. Everyone would be able to pay their debts with pocket change.

And fortunately it is quite simple to do this today. All that the prezs and prime ministers and heads of state have to do is instruct their respective central banks to put 6 or 7 zeroes between the integer and the decimal point in every account. Easy stuff! One quick stroke of the keyboard and we're done.

Or like an expert like you said: just change green money for brown or yellar money. That should fool the creditors!

But what actually is amusing is that all Disneyland Economists (Marxists, Keynesians, Friedmanians, Austrians, etc.) are experts in artificial economics, but complete brutes when it comes to Mother Nature's economy.

See   www.youstupidrelativist.com   Natural Economics.

Political science: How Admixture Became the Law of the Land
by

"Cowabunga" Bill Gaede

The hostile takeover

For professional paleontologists and lay enthusiasts, the extinction of our cousins, the Neanderthals, is an issue of great interest... or perhaps I should say "was." And whether the Neanderthals were in fact our cousins, is also a settled matter and now a thing of the past. The official version is no longer subject to challenge. The case is closed. All such issues have now been laid to rest and we don't need to think or worry about them anymore. And I suppose that, in a strange way, you could argue that that's somewhat of a relief. Decades of incessant bickering have finally and forevermore come to an end.

What happened in the last 20 years is that gold-digging geneticists gradually took over the Neanderthal extinction market of Paleontology. The spectacled geeks sensed their moment and went in for the kill. They ended up deciding the cause of the extinction of the Neanderthals for all of us.

What did these opportunists decree?

Led by crusader Svante Paabo, the geneticists at the Max Planck Institute for Evolutionary Anthropology in Leipzig, Germany, concluded that the Neanderthals were not such distant cousins after all. The Neanderthals were much closer relatives, more like our brothers and sisters.

The researchers found unimpeachable DNA evidence that the Neanderthals became extinct as a result of incest. It was intermarriage with the migrating Cromagnons – our forefathers and Neanderthal siblings – that did them in. And even that has been reduced to a figure of speech. If the geneticists have their way, it was more like "inbreeding."

The verdict of the geneticists is that the Neanderthals are still around. The Neanderthals are *us*!

A little bit of Paleo History

Until the end of the 20th Century, most paleontologists believed that either humans killed off the Neanderthals or the climate changed so abruptly that our "brethren" had no time to adapt. Climate change also involved humans, not because we caused it, but because we out competed the Neanderthals in the forbidding environment.

Our ancestors cornered the market for resources or had more children, or perhaps their more advanced tools and weapons gave them an edge over the backward brutes.

But a revolution was brewing among the rank and file of Paleolithic Paleontology. Frustrated with existing theories, the upwardly

How the Neanderthals Disappeared provides an alternative mechanism to explain the extinction of a single species: what is known as a background extinction. The most popular background extinction, of course, is the extinction of our cousins: the Neanderthals. How did Mother Nature single them out and leave the remaining animals in the vicinity alive? Was it a sudden climate change that our cousins couldn't handle?

The Paleontology Establishment and the geneticists will have you believe that humans had something to do with this measurable event, specifically that we interbred with this "Woolly Man." This book proposes, instead, that the Neanderthals went extinct like all other species that die all alone and without external help. They went extinct when they lost genetic diversity... as all species must after thousands of years of inbreeding. What kills a species in what is known as a background extinction is that its population pyramid finally overturns.

The corollary is that humans are on the same unavoidable path.

188 pages, paperback, ISBN 978-0-9704960-1-0, ViNi (2018)

Get the book... but be sure to watch the Neanderthal Series- Against Admixture on YouTube first..

https://www.youtube.com/watch?v=rMOJnkWjp5Q&list=PLwAqFWZDUpR8j0xcwT3NEAAM9r3clA6po

Raw Cabbages

JOIN THE DARK SIDE

Raw Cabbages

listen to new demos on SoundCloud.

check us out on iTunes.

NEW ALBUM COMING SOON!

Mobile paleontologists opened up a venue known as Admixture through which the new breeds of graduates could seep in and carve a name for themselves. Admixture holds that humans neither killed nor out competed the Neanderthals. Our forefathers drove them to extinction through incessant sexual intercourse. The smaller Neanderthal population simply dissolved into ours in the background over the years.

As occurs with most revolutions, at first the leaders had little success. What was missing was evidence. But what evidence do you look for to prove interbreeding? And how strong would the case be assuming something showed up? What Admixture theorists such as Milford Wolpoff and Erik Trinkaus did was produce evidence in the form of hybrids: remains deemed to have both Neanderthal and Cromagnon features. The search for half-breeds was so patently prejudiced that every specimen they inspected turned out to be a blend of Neanderthal and human. It got so bad that many experts could no longer unambiguously identify a pure bred Neanderthal through a bone that wasn't a skull. Of course, most old-guard paleontologists simply laughed at them and insisted on war and competition. But he who laughs last... ends up being the new peer-reviewer. The revolution was merely in its infancy.

At the turn of the century, things began to change and the tide turned in favor of the rebels. What tilted the scales was the invasion of the Paleo Guild by gangs of geneticists. Svante Paabo and his band of secret Admixture sympathizers got into the game. Paabo became the head of the Department of Genetics in 1997 and began his investigation of Neanderthal DNA. He started by wondering whether humans and Neanderthals shared mitochondrial DNA (mtDNA) The race was on against other genetic groups around the world to determine how much – if any – Neanderthal mtDNA we had in us today. Mitochndrial DNA is inherited solely from mothers. If the geneticists could determine how much of it we shared with the Neanderthals, they could do a regression analysis and more or less establish when the two species began to mate. The initial results in Genetics were very discouraging. Test after test of mitochondrial DNA (mtDNA) turned up negative. One study after another showed that the Neanderthals were a different species, meaning that we could not have successfully interbred with them. The conclusion was inescapable. If contemporary humans have no Neanderthal mtDNA, we had no Neanderthal mothers. The tests summarily outlawed Admixture.

To make matters bleaker for the Admixture lobby, a parallel study showed that human males have no Neanderthal y-chromosomes. The 'Y' is only passed on from father to son. This meant that human males never had any Neanderthal fathers either.

So... let's see... No Neanderthal mothers? No Neanderthal fathers? You would think that any defense attorney would take the case against Admixture with his eyes closed. Surprisingly, Admixture didn't die there. Instead, Admixture paleontologists kept the flame going by excusing the glaring absence of genetic markers with lame arguments. This gave their counterparts in Genetics a little breathing room to make the key breakthrough. The geneticists analyzed Neanderthal nuclear DNA (nDNA) and found a 2% match between humans and Neanderthals. This insignificant molehill was repackaged and sold to the public as a mountain. Was this tiny amount due to:
(a) the two species having a common ancestor in the distant past – some 500,000 years ago...
(b) or to interbreeding – some 50,000 years ago?

If the paleontologists went with option (a), Admixture died a sudden death. All geneticists researching Neanderthal extinction would instantly have to look for new jobs. What would there be to study if Neanderthals had no sexual relations with us... or worse... if the two species never even met after they parted from a common ancestor...? We could not blame them for diseases or thank them for antibodies. So what if a Neanderthal had such or such disease if he never passed it onto us? What would be the relevance? So who would invest in Neanderthal research seeking evidence of genetic diseases and antibodies? But three years before the Neanderthal genome was completed, foregone conclusions began to appear as shadows on the horizon. Times Magazine gave them away. Times named Svante Paabo one of the 100 most influential individuals on Earth. Now why would Times enter Paabo's name on the list if not because the editors realized that Paabo's monopoly gave him immeasurable power! Times realized that Paabo was in a unique position to judge the Neanderthal extinction issue all by himself. Talk about premonition!

And wouldn't you know it? The prediction materialized and came true. Another Cinderella story of success! People like that! A new celebrity to ask autographs from! Paabo and his team insinuated in their paper to their colleagues and to the paleontologists that *perhaps*, *maybe*, just *possibly*, there *could* have been, *perchance*, some isolated, casual interbreeding. Of course, the public was told something else. The popularization mags told the masses without reservations that interbreeding had been *proven*. No geneticist involved in the Genome Project denied the "exaggeration" or called the editors to retract their boasts. Quite the contrary! In interviews and public dissertations such as Ted Talk, the Admixture geneticists confirmed and reconfirmed, one right after another, the sweeping conclusion that the 2% commonality in nDNA pointed to interbreeding and nothing else. The revolution was over. Admixture had won the day. The war and climate factions were run out of town. They no longer call the shots. It was Paabo who made it happen.

The Admixture Revolution resulted in a complete routing of mainstream theories. The geneticists hijacked the Neanderthal extinction debate and took it along an irrelevant tangent. The discussion would no longer be about how the Neanderthals disappeared. After the Neanderthal Genome was finished, the debate would henceforth be not about why they're not around anymore. Now it would be about how much Neanderthal blood you have in you.

Where do we stand today?

Today, Admixture is the law of the land. You can't challenge the wisdom for fear of being labeled a heretic… or worse… a crank. You don't dare ask a paleontologist, "How did the Neanderthals disappear?" This question would show that you are simply out of touch. You might offend him by simply asking the question. No. The only question allowed is, "How much Neanderthal blood do you think I have in me?" His might smile and politely redirect you to a Neanderthal geneticist. "We don't do that kind of work in Paleontology. We just gather bones in the field and send them to the geneticists for analysis."

For some reason it always ends up like this in mainstream "science." The evolution of debate begins with honest research and ends in cutthroat politics. Once a theory gains the upper hand, there is a concerted effort to silence the opposition. This is not a deliberate conspiracy. It results from a Smithsonian "invisible hand" process: each individual pursuing his own self-interest. It is a strange mixture of theories and beliefs, careers and funds, power and authority, wanting to win the debate and silencing enemies. It occurred in Christianity after the Council of Nicaea in 325. It occurred in Physics after the 5th Solvay Conference in 1927. And it occurred after the results of the Neanderthal Genome project was finalized in 2010. It's human nature.

What are the merits of Admixture?

Is it possible to prove interbreeding with genetic data?

The answer is a definite and unambiguous NO! Interbreeding is an OPINION! It is the personal interpretation that a geneticist offers for the data he crunched. If you believe his word, he has "proven" the theory to YOU. YOU bought it. And usually this works on the basis of authority. People would rather have an "expert" tell them what to believe than to do tedious research and entertain alternatives. It's also more politically correct. As a bonus, you are not treated like a crackpot in social gatherings.

Did the Neanderthals really disappear by intermarriage… I mean… "in"-breeding? Did they dissolve into our much larger DNA pool?

Well, did the Woolly Mammoths disappear when they mated with the African or Asian Elephants? Or was it with the Mastodons?

A more fundamental fatal flaw with Admixture is that Neanderthals would have driven their parents – Homo heidelbergensis – to extinction also through interbreeding. And Heidelberg would have driven Erectus to extinction through interbreeding. And we can take the mechanism all the way to the chimps if it weren't for the fact that these hominids are still around.

Does Mother Nature use interbreeding to get rid of species? Is this the mechanism that causes species to disappear in the background?

If not, you should treat whatever comes out of the Max Planck Institute of Evolutionary Anthropology as nothing more than propaganda that enormously helps channel funds to the new lords of Paleontology.

Admixture Theory: the one on the left interbred with the one on the right.

Utopian Worlds
A curious experiment
By
Bill Gaede

Imagine that there is a group of herbivores in an immense, desolate region of the planet. You may think of them as camels living in the Sahara. The scene is that space is infinite and food is scarce. Will these animals multiply like rabbits? Will the entire mass of the herd exceed the mass of the available food supply? Now imagine a group of animals stuffed in a physically cramped enclosure, but with access to an unlimited amount of food. Imagine them as parakeets in a large cage being fed by a zookeeper. Space is scarce. Food is infinite. Will these birds multiply exponentially until the prison bars break? These two thought experiments point to the most relevant delimiters of what is known as *carrying capacity*: food and space.

Camels in an endless, food-scarce desert

Well-fed parakeets trapped in a cage

Let's now try one more thought experiment to zero in on the concept of *carrying capacity*. Let's assume that you take 4 pairs of healthy mice, male and female, place them in a large pen from which they can't escape, and feed them all the food they want like in the case of the parakeets. The males seek out a mate (hopefully of the opposite sex) and begin to build their nests (for which you provide abundant material as well). What do you think would actually happen in the following months?

From 1947 to 1987, psychologist John Calhoun performed a series of carrying capacity experiments like the one just described. His goal was to study the effects of density on the *behaviors* of rodents, speculating that he could extrapolate the results to humans. In his quintessential experiment known as the Utopian Universe 25, Calhoun placed four healthy couples of mice in an enclosure. He supplied them unlimited amounts of food and allowed them to reproduce. After about a year and a half the population stabilized at around 2000. Thereafter, the mice stopped reproducing. Throughout this period, the social structure increasingly deteriorated until the colony finally went extinct.

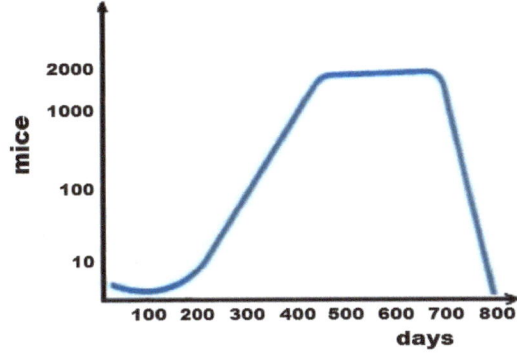

Calhoun's Mouse Utopian Universe

Calhoun investigated the mice from a psychological perspective and concluded that the animals could only withstand a certain amount of social interaction. With so many mice trapped in such a tight space the number of minute encounters was so great that the mice couldn't handle the "rat race." It overwhelmed their brains. Autism and homosexuality (including asexuality) developed in the last generations and the rodents eventually stopped having offspring. Calhoun summarized his observations in a disturbing one-liner: "They stopped being mice."

There were, however, a couple of invisible variables that Calhoun may have overlooked when explaining his results and which have an inordinate bearing on population dynamics. One is the Founder Effect and the other density-dependent birth rates (DDBR).

The Founder Effect

The Founder Effect occurs when a small group of individuals found a colony. After a few generations inbreeding results in low genetic variation. Physical and mental disorders develop and the last members forego procreation. Unsurprisingly, a small colony quite swiftly ends in extinction.

Calhoun's utopian experiments suffered the ravages of the Founder Effect. His subjects consisted of domesticated lab mice, a situation that already suggests less than optimal starting material. Thousands of years of evolution and population bottlenecks along the way have already diluted the gene pool of mice around the world. A population bottleneck occurs when a disease, accident or other factor eliminates a significant proportion of individuals from the total. Hence, whatever group of mice we select for a study already has quite an advanced level of genetic uniformity, more so if they are lab bred. Researchers that play around with fruit flies and guinea pigs in their labs usually fail to factor the number of years the species has been around. When did Mother Nature spawn the species? How many founder effects and population bottlenecks did it experience?

The founding fathers of Calhoun's colonies already had their genes washed up when he placed them in the pen, and it didn't take long for inbreeding to do the rest. After the fourth or fifth generation, the mice were coming out retarded, autistic, and homosexual. He dubbed these socially isolated mice "the beautiful ones" because they were perfect physical specimens of mice, yet they no longer acted like their free-roaming wild brethren. Was it the population density that somehow drove the mice out of ther minds? Or were they born that way? Nature or nurture?

Density-Dependent Birth Rates

From a biological and ecological perspective, what Calhoun did is confirm the law of *density-dependent birth rates*. Contrary to what Malthus and Darwin believed no species of plant or animal multiplies exponentially forever. Animals do not produce offspring mindlessly as most people believe, but rather adjust their reproduction rates as their numbers increase. Density could mean the number of individuals with respect to either space or to food. Calhoun's mice would have stopped reproducing or lowered their reproduction rates just as well in response to the amount of food. It is a myth that animals put out litter irresponsibly and then watch starvation take its toll.

Several studies have shown that animals adjust their birth rates in response to density. In the wild, density is typically a function of the available quantity of food.

Random papers arguing that animals produce fewer offspring at higher densities:

Fowler: Density Dependence as Related to Life History Strategy

Rubenstein: Individual Variation and Competition in the Everglades Pygmy Sunfish

Southern: The natural control of a population of Tawny owls (STRIX ALUCO)

Bonacic et al.: Density dependence in the camelid VICUGNA VICUGNA: the recovery of a protected population in Chile

Stewart et al.: Density-dependent effects on physical condition and reproduction in North American elk: an experimental test

Even humans are bound by DDBR...

Lutz et al.: Human Fertility Declines with Higher Population Density

Plant populations have also been confirmed to adjust reproduction in response to density.

Random papers arguing that plants produce fewer seeds at higher densities:

Harper: maize (Approaches to the study of plant competition)

Watkinson: dune plants (The demography of a sand dune annual: Vulpia Fasciculata)

Clay and Shaw: elf orpine (An experimental demonstration of density-dependent reproduction in a natural population of DIAMORPHA SMALLII, a rare annual)

Schmitt, et al.: jewelweeds (Density-dependent flowering phenology, outcrossing, and reproduction in IMPATIENS CAPENSIS)

The popular notion that living entities other than humans are mere automatons, machines that function robotically and spit out progeny incessantly without regard to food or space, is a myth. A tigress which is feeling hunger pains and realizes she has trouble finding food does not bring a kitty into this world. She first takes care of Number One: eating before sex. Animals do in fact practice family planning! They do it the natural way: instinctively. If Calhoun had limited the amount of food, the mice would have adjusted their numbers before suffering the ravages of spatial density (not to mention the fact that wild mice also struggle daily to find food).

What Calhoun actually showed in his experiments – if anything – is still heatedly debated among the handful of psychologists who are familiar with his experiments. Do they suggest that density affects the minds of mammals through intolerable amounts of social interaction? If so, do humans have a greater tolerance? Was low genetic diversity the critical parameter that ultimately affected the overall mental health of the colony? We certainly have more mental disorders in our contemporary world. There are visibly more autists, schizophrenics, homosexuals, transexuals, and asexuals all around us today. Is this the result of overcrowding? Are these individuals coming out of the closets or out of the wombs? Are they the human versions of Calhoun's 'beautiful ones'? Or were the mouse experiments a portent of things to come: of what will happen to humans as they continue to crowd into cities?

What caused the human population pyramid to begin to overturn?

The human population pyramid began to overturn in 1963. Since then the global population growth rate has been *steadily* declining. Demographers can now pretty much predict when the global population will peak. If their predictions are more or less on the money, we will attain this historical milestone during the 21st Century. Yes! Please memorize it! For the first time in the history of modern humans – spanning some 200 ky now if we are to believe geneticists and paleontologists – the birth rate is GRADUALLY and PREDICTABLY falling and we can't do anything about it. Neither intelligence nor technology can reverse this trend (*...any more than intelligence or technology can overcome the need to eat, sleep, breathe, or die*). We WILL attain Zero Population Growth this century no matter what!

What happened in 1963 is that our species crossed the line into old age: we started having fewer babies per family on the average. We became an *old species*.

United Nations demographers attribute this steady trend to a variety of reasons, including income level, social status, labor market participation of women, education, family planning, proactive NGO intervention, infant mortality rates, etc. Make no mistake. None of these factors addresses the root cause. The reason behind the global decline in birth rates is that people moved from the country to the cities. What increased dramatically in the last 50 years is the *urbanization*! We are still massively migrating from open spaces and suburbs to the densely packed metropolitan areas. Once people settle in urban areas, cultural and economic constraints do the rest. It is simply unimaginable for a young couple to have 10 children in a city apartment even if both the man and the woman came from traditionally large rural families and had the incomes (and the cultural background) to support so many offspring. We are boxing ourselves into invisible enclosures no differently than Calhoun's mice.

The transition to the urbs was an unavoidable process. In retrospect, it was predictable that the human economy would evolve and that the typical family would evolve with it. In the last 200 ky, we went from hunter-gathering to farming tomanufacturing to services. Each of these economic phases can readily be associated with a specific type of family. Hunter-gathering was the age of group marriages, agriculture was the age of patriarchy, manufacturing was the age of the nuclear family, and services is the final age consisting of single mothers, bachelors, and homosexuals. The typical "family" has gotten smaller over the centuries. We went from the many to the few. The prevailing economic order maintains a symbiotic relationship with the median family type.

Those eras are gone. They melted away and were steamrolled over by services right after World War II. Farmers left their lands and moved to the cities where they found jobs first in manufacturing and now in services. This transition is still in full swing across the planet in developing nations. We simply cannot and will not have marriages or children in the cities. Our demographic expansion will finally come to a gloomy end. There is no need for militant environmentalists, NGOs, and Neo-Malthusians to insist on family planning. It *will* happen!

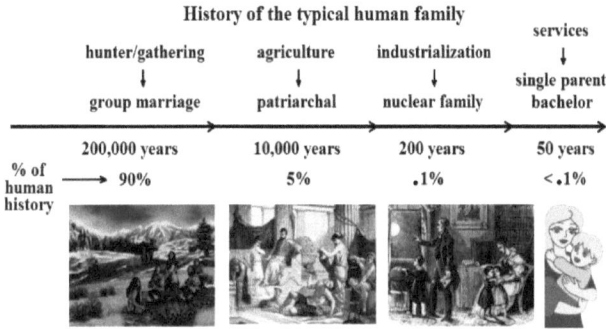

We should not be surprised, therefore, to find that humans ended up like Calhoun's mice: corralled into ever increasing densities. It was predictable that we would end up living in crowded cities and developing similar pre-extinction symptoms such as sexual and other mental disorders. Like the mice, we cannot escape the two variables that had a decisive bearing on his experiments: the Founder Effect and DDBR. Even our population curve is starting to look like theirs!

The typical S-curve that biologists and environmentalists illustrate is incomplete. In the long run, it continues until a species ends in extinction like Calhoun's mice

Mankind is blessed with intelligence, foresight, and technology, yet there was nothing we could have done to prevent or that we can do now to revert these trends. We have as much chance of going back to an earlier family type as we have of going back to hunter/gathering. We are far too technologically advanced to go back to a manual agriculture economy... which in turn means that the family will never again shift back to a patriarchal family type. And we are also not going back to a manufacturing economy or the nuclear family that served as its foundation.

typical population S-curve

Population curve of mice

Not only were these trends predictable, but they irreversibly lead to extinction. Like all other animals, humans are subject to Mother Nature's laws. We have no power to avoid density-dependent birth rates and eventual extinction to which these conduce... any more than Calhoun's mice could. As our population density increases in the cities, we will have fewer children. The fewer babies, the older the population gets. The demographic pyramid overturns permanently never to reverse itself again.

Overturning of the population pyramid

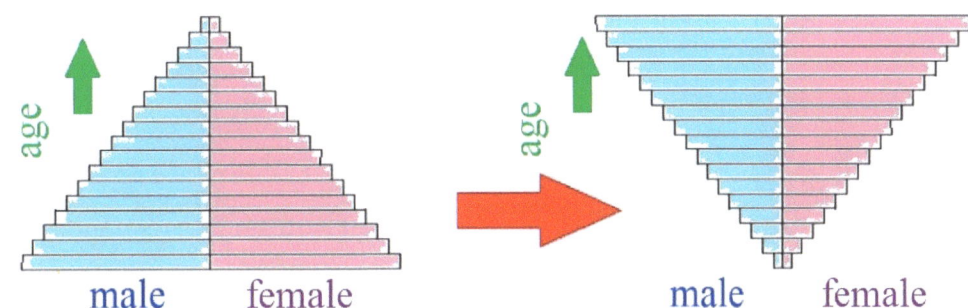

Population pyramid inversion leads to extinction

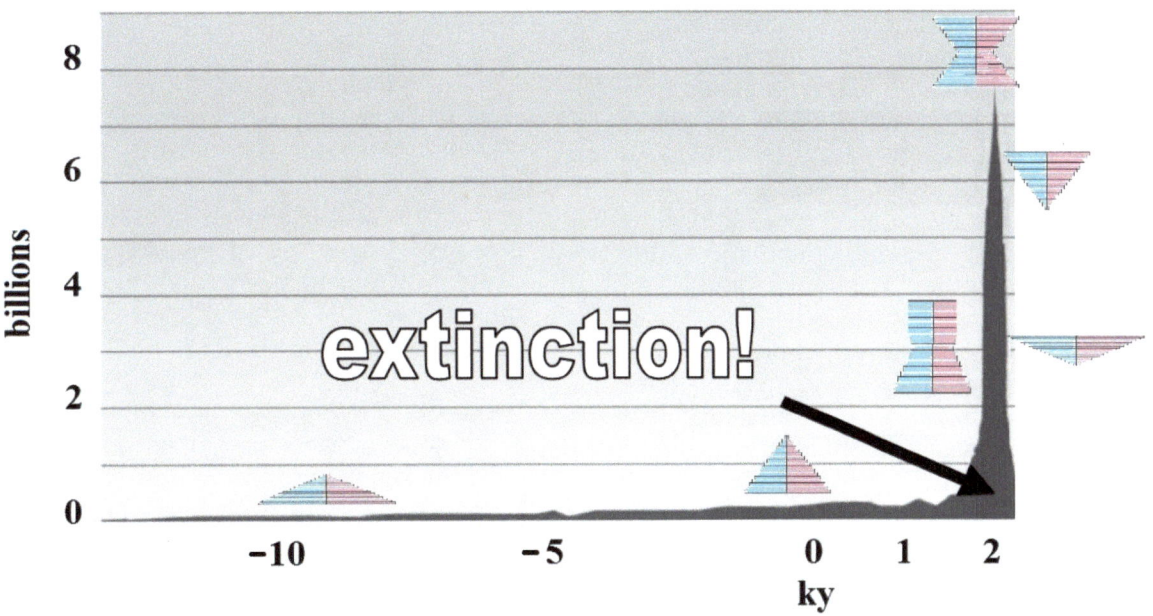

Bugworld

By
Monk E. Mind

Bill and Thomas were colleagues, and the best of friends, but that didn't mean they always agreed on things. They had more than one heated discussion on what Bill called "natural economics" as it applied to man. Thomas thought that man was different. At least, he used to think that. Now he wasn't so sure.

"There are at least five things that are spelling doom for man," Bill Randall said. Increasingly less women are getting married, and even less are having babies. We've been intermarrying for thousands of years, with city life accelerated the process by which our genes have been washed out. Man has become "Homo hillbillycus," and, because of the lack of genetic variation, has limited his adaptability and survivability. We've pissed in the gene pool. There is less genetic diversity with man than our closest relatives, the chimpanzees. Population pyramids have overturned in nearly every country, and soon the entire world will be inverted. Service based economy is the last category, and technology is killing jobs. Governments are fueling their country's economies by creating more and more money. All countries are becoming more and more dependent on other country's exports.

"Right now about 3% of the population provides the food for the rest of us. It is mostly these behemoth agri-corps. If they "go under" we starve. This could easily happen with a global collapse of man's artificial economy. Once there is no profit, there will be no incentive for ADM, Nestle, and Bunge to produce food for you. Birth rates continue to decline, and once we are at zero population growth, that's it for man. But that will never happen because the collapse of the artificial economy will preempt naturally occurring zero population growth."

"Your theory is great for the rest of the animal kingdom," Thomas replied. "Where things go wrong is when it's applied to man. I argue that you are misdiagnosing the cause of unemployment. It has nothing to do with the progression of a developing economy through your defined economic stages. More developed economies outsource types of production like manufacturing because it's cheaper to pay someone a dollar a day to work in a factory than it is to pay some westerner who is used to a higher standard of living. Also, a developing economy, like China, which is experiencing its own industrial revolution, needs to produce the means of production necessary to elevate their population away from menial labor. All it takes is capital accumulation and the division of labor to achieve this."

Bill strongly disagreed. "Division of labor is NOT unique to man. One of the hallmarks of eusociality is division of labor: some stay home and care for the nest, some forage, some care for the young, and some go to war. Eusocial insects have co-evolved with plants and animals accumulating "capital resources" for millions and millions of years.

"An example of co-evolution can be seen in ants with aphids who hold back their excrement until ant antennae tickle them. It is a source of food for the ants that raise them, clean them, and feed them.

"But while eusociality makes the colony more efficient, prolonging colony life, primitive eusociality does not prevent extinction of a species, and it does not guaranty a colony will not revert. Halictid bees, for instance, have developed eusociality probably three times, yet have reverted from primitive eusociality to solitary behavior a dozen times.

"Humans are only loosely considered eusocial, and more closely resemble primitive eusocial insects than the highly eusocial ones. But human eusociality quite likely is directly behind much of the mass extinction of species that is occurring right now because we have not had the time to co-evolve with other species. We have gotten good at trashing our environment, and eating anything that moves."

"The division of labor spanning the entire globe is unique to man. Capital accumulation over thousands of years is unique to man. Man is unique in that procreation depends upon how they expect their standard of living to be in the future with children, rather than just considering their present condition," Thomas argued.

"Not so!" responded Bill. "Eusocial insects, and other animals can be found all over the globe. Insects by shear numbers and mass dominate the entire planet, save the poles.

"And if you think birth control is unique to man, you better think again. The queen ant, for example, can produce female drones, or male mates, at will. The decision of birth for eusocial insects has to do with the health of the colony. It's one reason why they are succeeding where we are failing. However, if one colony collapses, it doesn't insure a global collapse like man's artificial global economy will. If this was special to man, it is nothing to brag about, as it is that which is speeding up his demise, not strengthening his survivability.

"While eusocial insects dominate the invertebrates, and man dominates the vertebrates, man hasn't had time to learn how to live with the rest of earth's creatures (though insects have). Man has taken far more than he has returned, and is himself the driving force behind the sixth mass extinction. AND, it will be man that goes extinct long before the last of the eusocial insects. Then they will be consumed by single celled organisms before the sun cools, and everything else on earth dies.

"So the cycle of background and mass extinction continues as a natural part of "assembly, disassembly, and reassembly of all objects" as you like to say.

"But, it is not unique to man or insects. Plants globally clean the air, provide oxygen, and aerate and replenish the soil. They also perform and share other tasks to varying degrees. Whales travel the world, and manage their capital resources quite well, until man gets in the way. The Arctic Tern breeds in the Arctic, but is found everywhere.

"Insects have accumulated capital over millions and millions of years, and it did not stop thousands of species from going extinct. So Man is not special in this regard. There is no reason to believe that man CAN conceive of what his standard of living will be like years down the line or that it will, or has helped man forestall the overturning of the population pyramid."

"What," protested Thomas, "plants do not perform labor, there are chemical reactions that perform those functions. Only animals can perform labor, that is, the directing of efforts and actions towards certain ends.

"Insects have absolutely not accumulated capital over millions and millions of years! Capital is goods that are used in production. Man had to start somewhere. From the first guy who picked up a stick and sharpened it to make a spear, all the way down the line to the formation of plows and factories and machines, it has been accumulating for thousands of years. What goods do insects have that they have passed on from one generation to the next allowing them to develop more and more means of production? What, a stick? For millions of years?

"Man doesn't perform labor, chemical reactions do," Bill responded sarcastically.

"I already gave two examples of goods passed on from generation to generation. Here they are again. Leaf cutter ants grow their own fungus. It is found nowhere else but in their nests. Many species of ants raise aphids which became simbiotes. This has been going on for millions of years.

"Animals also plan for the future, such as a squirrel hiding away his cache of nuts for the winter, or a dog burying his bones.

"This capital does not accumulate and develop across those years," argued Thomas. "They can grow fungus or aphids, great! What is it that they pass on from generation to generation that continually develops, and builds for millions of years?"

Bill answered, "They don't need to accumulate anything but the ability to grow fungus, and raise aphids. And man can not accumulate food for millions of years either. Food spoils. What is passed on is the ability to cultivate, and to raise cattle or aphids."

"The difference between man and insects is man continually produces capital, and recombines it to form higher order capital goods. This process has gone on for thousands of years from the guy who sharpened the first stick to the guy who used machines to produce a rifle, or whatever. That long chain of actions was necessary to get us where we are.

"Man's superior ability to engage in the division of labor on a global scale, and capital accumulation over the course of thousands of years is a testament to his ability to conceive of future possibilities, and devise the means to obtain them. Man vastly outpaces the insect's ability to conceive of future possibilities and to conceive of the means which might allow them to achieve their ends."

"Not quite, Thomas," Bill said, "only a few humans can rationalize the extinction process. Only a handful of humans on Earth have the critical reasoning capacity to rationalize without contradictions. Not only insects, but every species has developed technology. Division of labor isn't restricted to man and insect either.

There are at least two types of division of labor in clonal plants, and even within simple multicellular organisms individual cells differentiate and specialize which requires communication between cells. One cell can secrete a substance causing a neighboring cell to change its physiological activity and biochemical traits, and it can develop new functions.

"So man isn't unique in respect to division of labor even when considering plants and simple multicellular organisms. It's just a matter of degree. And some research is also showing that plants can conceive, feel, smell, warn neighbors, and mount a defense against predators. Plants even have what is considered a type of memory.

"There are literally hundreds of studies...... So it's worth considering that plants, although obviously different than man and other animals, have a degree of intelligence, have their own sensory system allowing them to respond to danger, can respond to that danger, and can communicate it to other plants."

"All plant and animals evolved to survive, some are better at it than others. Remember that you are trying to make the case that man is so special that he can outwit Mother Nature. Man is better at many things. He is not better at survival. By any means!

"In fact, the singled cell organisms are the best survivalists. They were the first here, and they will be the last to go. So much for man's artificial economy!

"And think about this," Bill continued, "If matter and motion are eternal, and man is special, this Universe would be filled with intelligent humanoid life everywhere. Every planet in this and every other solar system would be covered with intelligent beings regardless of atmosphere, etc., because technology fixes everything! But this isn't the case in spite of man being special for eternity!"

"O.K., I see your point," Thomas acknowledged, "but what catastrophic economic event could possibly stop man from producing food?"

Bill answered, "It would only take an economic collapse where money has no value to stop food production."

"What is "artificial" about man's economy? The fact that he produces and trades in goods other than food?"

"Yes. It was the trade of goods that led to the idea of money and man's artificial economy. Humans removed themselves from nature's world when they did this. The currency of the natural economy is food, but you can't eat money."

"What will prevent already existing resources from being reallocated to new owners for continued use?"

"Time. There won't be enough of it. It takes time to grow crops, and raise chickens, then get those to the grocery store."

"What will interrupt food production to the extent that all the existing resources are wiped out?"

"Food is grown to fit the current population demands. If food production stops, the shelves go bare within days, but it takes months to raise crops and livestock, and then deliver that to the market. All it will take for food production to stop is if there is no profit in it for the ones who are providing it."

"You said that the population pyramid has overturned. How is it already overturned?"

"Just Google "is the world population aging?" Guess what? It is! Rapidly. The population stabilized in 1968. That's when the theory in question proposes the population pyramid overturned. It started when birth rates starting going down. Here's an interesting factoid for you: Last year in the US more white people died than were born. Anyways, when we hit zero population growth there will be no turning back.

"And the Economic pyramid?"

"Yes. The currency of the natural economy is food. If food sources dry up...mass extinction. When small family farmers began walking off the farm, and agri-biz took over, that spelled the doom for mankind right there."

"The crux of the issue is Density Dependent Birth Rates for man. What will cause human females to stop procreating? If food production won't cease, why will they stop making babies?"

"These are two separate issues. We've talked about this before. Calhoune's mice had all the food, shelter, etc., they wanted. Density Dependent Birth Rates has to with the population pyramid. Although individuals have free will, it is the group that limits the choices for the individuals.

"What about unemployment? I argue that you are misdiagnosing the cause of unemployment. It has nothing to do with the progression of a developing economy through your defined economic stages."

"What comes after the service industry? The hobby industry? I suggest it is the unemployment 'industry' and the last stage of the artificial economy before reverting to the natural economy where food is the only currency and everyone becomes a hunter gatherer."

"Your idea, if I understand correctly, is that the entire global economy, in general, is tending towards unemployment because the robots are taking all the jobs."

"No, I think it's more that we have everything we need, and can't think of the next stage. Things have cycled down. Hunter gather-100,000 years, agricultural society-10,000, industrial society-150 years, service industry-50?, unemployment age?, Age of the Hobby?

"I'm arguing that unemployment in the long term is not caused by technology, it's almost entirely caused by the government."

"Well, if this is the case, then we are almost certainly doomed, for that reason, if no other.

"Now, leaf-cutter ants. Those guys are smart. They help out with the environment by trimming the upper arborial region of the forest. They raise their own fungus on the leaf matter they bring into their nests. So, they created their own food source. There is no fungus like this anywhere else on earth.

"This is just one of many examples of how insects work with the environment, and how they have co-evolved with it, insuring plenty of resources for a long life and their dominion over earth."

"We'll have to continue this conversation later, Bill, I need to go home and get some rest. I'll think about it some more, and we can discuss this idea about the 'last stage of the artificial economy' some other time."

...

Another weekend, another 12 pack, Bill and Thomas were at it again.

"The one and only job category is called entrepreneurship, that is, the management and directing of resources," said Thomas Mann.

Bill Randall responded, "And again, what is it that the entrepreneurs will come up with after services? Entrepreneurs have been around since the beginning of man. They have been in every culture and every age. It didn't stop the trend towards unemployment.

"Concepts such as capital accumulation, or higher order capital goods are not special in a NATURAL ECONOMY, and will not make any

Are Mainstream Science Magazines Really Science?

THE RATIONAL SCIENTIST

APRIL 2018

NO! It's NOT science, Mildred!
In this issue you'll Discover why.
It might be popular but it's NOT Science!
It might be American, but it's not Scientific.

Energy, Waves, Mass, Fields: These are Concepts not Objects. Learn the difference and never be fooled again!

difference when food is the currency. Man's artificial economy can not replace nature's economy indefinitely. Once the human population reaches zero population growth the artificial economy will very quickly revert to the natural economy. Sooner than that if the global economy collapses.

"Thing is, even if you can show how man is special, you'll still have to show how that unique special quality or qualities will allow him to overcome the mechanisms of extinction."

"Human action is characterized by means and ends. All conscious, deliberate action involves a directed effort to achieve a certain state of affairs (ends) via certain means. Insects, and the rest of the animal kingdom, are merely reacting to stimuli according to their nature. Man vastly outpaces their ability to conceive of future possibilities, and to conceive of the means which might allow them to achieve their ends. This is how we developed technology, and a society that spans the entire planet. Man IS SPECIAL in this regard, Bill...but anyways, gold and silver will not lose their value, they will regain their status as the world's reserve currency," Thomas opined.

"Thomas, in an economic collapse with starvation, only food has value. You can't eat gold or silver."

"But, Bill, all the money in the world will not lose its value simultaneously. Especially not gold and silver, and as such, there will STILL be an incentive for agri-corps to produce food."

"Sure it will. Just like dominos going down. Knock over the US dollar, and the rest will fall. There won't be enough time for any of what you propose. All the incentives in the world won't stop the collapse from happening. After the world economy collapses because currency has no value we revert to the natural economy. In a natural economy it is impossible to trade for food because the only thing of value IS food. Animals don't trade. There is nothing to exchange, just hunting and gathering.

"We see that nearly every species that has ever lived has gone extinct. We observe a reality of eternal motion; of assembly, disassembly, and reassembly, and we reasonably assume that everything will come apart eventually.

"The amphibians of the Carboniferous died because of lack of food, and were replaced by the Permian synapsids which died because of lack of food, and were replaced by the Triassic archosaurs which died because of lack of food, who were replaced by the Jurassic dinos, which died because of lack off food, which were replaced by the dinos of the Cretaceous which died because of lack of food, which were replaced by the Cenozoic mammals…

"BECAUSE the last animals at the end of a long dynasty die of starvation we have invented pretty names to refer to the geological eras. The Carboniferous is known as the Age of Ferns. The Jurassic is known as the Age of Cycads. And our very own mammalian era is the Age of Angiosperm or Flowering Plants.

"Man's intelligence can not come up with anything that will help him avoid the overturning of the population pyramid and the economic pyramid. Man has already exceeded his carrying capacity. We will experience die-off and then die-out. This is what happened to the passenger pigeon, Easter Island inhabitants and the Dodo.

"Take a look a yeast cells in a wine vat. A very successful species that eat up the nutrients and increase their population; in a few weeks the 'pollution' they produce (alcohol and CO_2) fills up the environment and then they experience die-off and extinction. It is no different than what happened with the reindeer on St. Mathew Island where…"

"O.K., O.K., enough! I think I get it" Thomas conceded. "It's just a very hard pill to swallow, Bill. But there's one thing that doesn't add up. Your main argument is that urbanization and stress put a downward pressure on birth rates. Yet Africa has a higher population density, and vastly worse food problems than America, yet their birth rates are significantly higher. Africa defies your theory of Density Dependent Birth Rates. Human birth rates have little to do with population density, and everything to do with the economic burden or surplus of having a child. Birth rates decrease as economic development increases."

"Yes, and economic development increases because agrarian and industrial based societies are moving towards the service industry based societies. There are global declining birth rates, *including Africa.* You have to compare Africa today to Africa of yesterday. The key issue here is that once the entire world is service based everything grinds to a halt. Nothing left but underemployment. So, once more I ask, what comes after the service age?"

"I don't know."

"And neither does anyone else."

The professors' conversations took place before Harold's Hoppers. They were two of only a handful of people on Earth that had a clue of the impact this would have as a catalyst, a last straw on the camels back, a tipping point in the overturning of the economic pyramid.

The two friends bought food and supplies, and a lot of beer before locking themselves up in Bill's apartment. They remained there until their end, which came without them even noticing, about three months later.

...

In the U.S., the Army, National Guard, and FEMA, lead by the Department of Homeland Security, couldn't quell the masses of rioting people. When people couldn't access their bank accounts and buy food for their families, they began breaking windows and climbing into storerooms to help themselves to whatever they could find. It wasn't just the devastating crop damage, or the 15 year drought that flipped over the economic pyramid. It was the unemployment. Not just the lack of jobs available in automated manufacturing and service-based economies, but also because of the lack of workers in the aging population. The population had slowly aged since the 1960's and each country eventually exceeded its carrying capacity as it reached zero population growth. The working young simply could not support their elder population.

The government kept borrowing from Social Security and printing fiat money. Medicare alone quickly exceeded the National Income. Debt held by the public as a percentage of the Gross Domestic Product rose to a hundred percent.

The financial crisis at the Treasury rapidly moved through the Federal Reserve and all the banks destroying the Main Street Economy. Businesses quit hiring, and unemployment rose to 40 percent. Stock market prices dropped, and 401K's became practically worthless. The value of the American dollar dropped as prices of everything went up. Lack of confidence triggered mass selling of the U.S. Dollar destroying it as the world's reserve currency because countries were wary of accepting it in trade. All American goods prices from clothing, to food, to gas went sky high. Social Security checks stopped being mailed out. Medicare and unemployment benefit payments ceased. Bank lending stopped, and immediate payment on loans was demanded. Money Market Funds tanked as global markets began to fail.

The US couldn't pay its debts missing its debt interest payments for the first time (publicly). It had happened before, but China, Japan, Germany, and Saudi Arabia helped cover it up because they had plenty of liquid assets to hide the debts, while the U.S. printed more fiat. Immediately, the government started dumping trillions of dollars of U.S. Treasury securities on the global financial markets causing hyperinflation and a ripple effect which eventually caused a depression the likes of which had never been seen before.

The Secretary of Treasury declared a force mejeure on debt service. When news of the wire was broadcast around the world, it was seen for what it was, a repudiation of debt.

This triggered the 6900 series of protocols. Special Forces were parachuted in around the country to protect the 15 FEMA and 12 Federal Reserve district banks, as well as, other federal installations.

The NASDAQ, SEC, and other regulatory agencies instituted emergency protocols, such as, "no more stop orders" to no avail. The Dow indices dropped 40% in value in one day.

News networks, such as CNN with their talking heads, informed the public overnight, and the dire situation began to sink in. Tokyo financial markets closed within a few hours of opening, and European markets didn't bother to open at all. The next day all financial exchanges in Europe were closed. The US tried to maintain liquidity, but the DOW lost the rest of its value. By the next day, the Federal Reserve ordered all markets closed, and declared the money worthless.

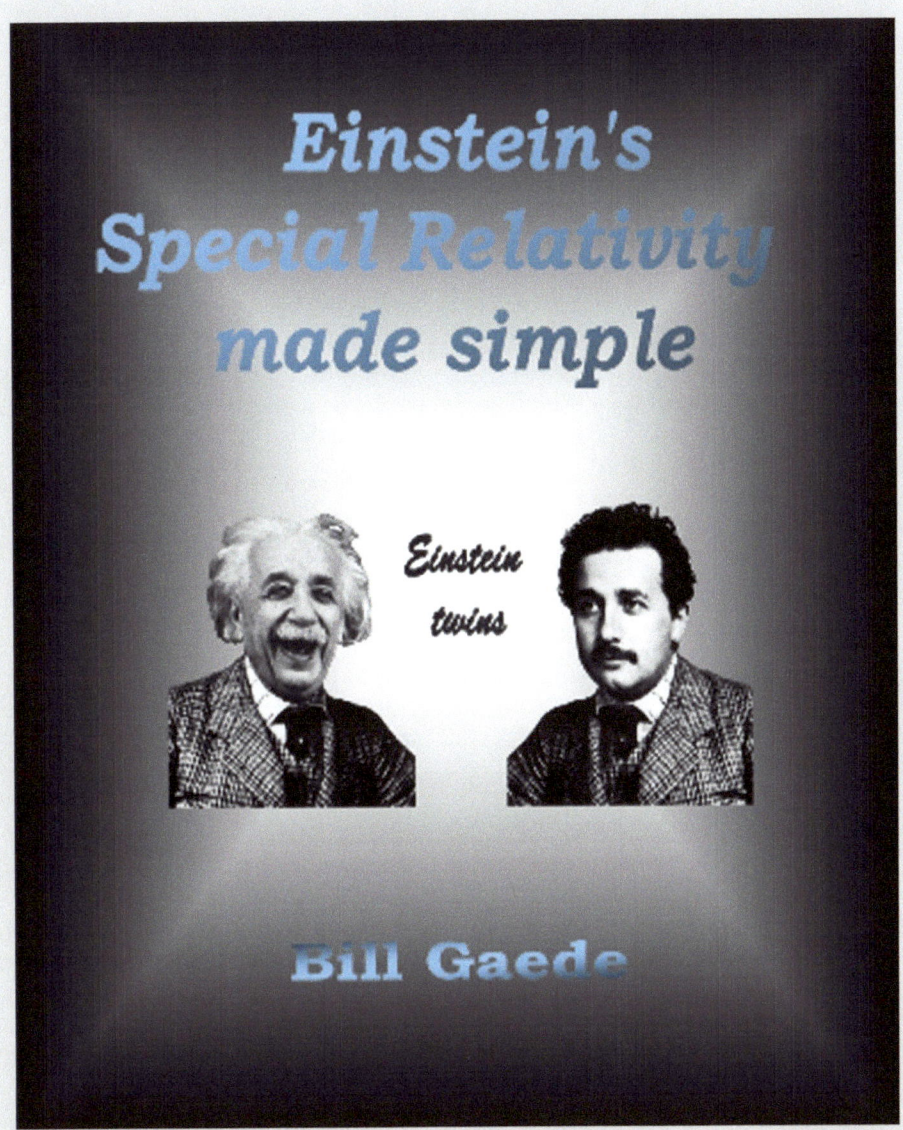

Special Relativity made simple is the ideal book for laymen with open minds who suspect the fantastic explanations they hear and read about from authoritative 'scientific' sources. Does it make sense to say that twin brothers could differ in their ages by 50 years? Is it rational to say that you can travel to the past or to the future?

Special Relativity made simple addresses these and similar questions and arms the average reader with arguments that enables them to challenge Einstein's theories on the Internet and in conferences.

To obtain a paperback, Paypal to bill@youstupidrelativist.com
USA/Canada   US $20.00 Plus $10 Shipping
Europe 20.00 Euros Plus 10 Euros Shipping

The President declared a state of martial law.

Because they were the largest holders of U.S. Treasury Debt, the fall in U.S. Government bond prices wiped out Japan's and China's reserves, and they threatened to retaliate militarily. One by one, trade stopped between countries resulting in global economic collapse.

In the United States, the straw that broke the camel's back came when the agriculture industry collapsed. When the three percent of the population that provided the food for the 97 could no longer make a profit, they walked off the job.

The grocery shelves were emptied in less than three days never to be filled again.

After the hedge fund and money market fund crisis and crash of 2008, the D.O.D. started looking into various scenarios of how global markets could threaten National security. The Army instituted Unified Quest. The Marines visited J.P. Morgan to study the economy. Its Chiefs of Staff were trained in handling riots in case martial law was needed to keep the order. Soldiers were trained in diplomacy and negotiation to be equipped to deal with angry and hungry citizens. It didn't help matters at all.

Peer to peer trading only worked temporarily and only in some local "sharing economies." When people are hungry, all they want is food. Bitcoins and other private currency was worthless. One can not eat gold, coins, or cash.

Although the U.S. Government had been the largest purchaser of freeze dried foods, and many people had purchased emergency shelter equipment and supplies for their bunkers, the overwhelming majority of people didn't have survival supplies or food enough for a week.

It was basically the same everywhere. Russia's 5,000 underground bunkers in Moscow, the European Union's seed bank, the millions of shelters, bunkers, caves, cabins, and survivalist caches couldn't save but a small percentage of the population.

Shoot to kill orders where given and for a while the 1.6 billion rounds of ammunition purchased by DHS were used against the American people. But in the end, tanks, troops, guns, and bullets couldn't stop the starving masses. Most cities shut down their utility departments and the lights went out. People lost their electricity, water, sewer, gas, internet, and cell phone services. Radios and TVs were dead, but the few people with battery powered radios communicated for a while. Panic struck the cities as people went crazy with hunger. Folks tried to get out of cities. Urbanites and suburbanites headed for the country's state parks, forests, woods, and wildernesses.

The highways were controlled by the military, but most interstate trucking was stopped.

There were plenty of guns, but not enough people to aim them at the roving gangs of desperate, hungry people. Most of the soldiers had families of their own and soon abandoned their posts to be with them.

Under DHS control, the Sheriff and Police departments deputized any one willing to carry arms, offering food to those who joined.

Grocery stores, gas stations, warehouses, and all sources of food and fuel were shut down. The Special 117th Assault Unit parachuted in and seized control of the Oklahoma City cattle yards as the military took control and scaling circles were implemented. First they had to secure the stocks of food, fuel, and distribution points, or the starving masses of people would get them. Anyone trying to steal was shot.

In the beginning, the economy converted to a basic one of bartering food for supplies and services. Persons with skills like farming, cooking, organizing, warehousing, doctoring, and diplomacy found work within the scaling circles as the government passed out ration cards in lieu of MREs and fuel. "Scaling circle scenarios" was the Department of Defense and FEMA's way of saying, "stop domestic resistance in the midst of an economic collapse" or "quell large-scale, unexpected civil disturbances." These scaling circles were supposed to be used to contain the spread of disaster and diminish its ability to affect areas outside of the circles.

Meals, ready to eat, also called military rations, were dropped to secure locations around the perimeters of the scaling circles, but few made it to within the cities. At first, military, National

Guard, and civil authorities, along with the "good old boys," kept the rations for themselves and their families, or doled them out to persons with gold or some kind of powerful influence.

Anyone who questioned the government was considered a "threat against government stability." The military, under the DOD's Army Modernization Strategy, considered any such person as a major threat against the government of the United States of America and they were dealt with accordingly. At first they were housed in makeshift prisons, but when food starting running low, they were taken out to a landfill, shot, burned, and buried with the trash.

That didn't last for long. When they finally understood there was not going to be a recovery from this economic collapse, and that the world had reverted to a natural economy where the currency was food, they abandoned their posts, took their families, and headed for the outlying areas.

Chaos ensued in the towns and cities of America. Identical scenarios played themselves out around the world.

Persons in good physical shape; campers, survivalists, military personal, and others with survival skills, moved out into the forests, parks, and rural areas.

Many people locked themselves in their houses and apartments with what small rations they had. After that ran out, they scoured the warehouses, stores, and pantries of every home, apartment, restaurant, or hotel, but without any success.

In cities and towns, friends turned on friends, neighbors on neighbors, and family members on family members as hunger drove them all insane. When they couldn't steal each other's food, because none was left, they killed each other and cannibalism was the only recourse to those who wished to remain alive.

Packs of wary cannibals kept each other at arms distance, while roaming the 'burbs in search of what they called Spam.

When the cities were emptied, they moved out into the rural areas, and then into the parks, woods, and forests in search of survivors.

...
"Hormel....Hormel... we know you're in there. Open up, we're your friends...don't you want to invite us to dinner? ha ha ha ha..."

The family huddled together trembling in fear as the inhuman pack of ravenous maniacs taunted them while trying to pry open the safe room door with a fire poker. It was the only thing standing between the Hardwicks and certain death.

"Spam, Spam, Spam, Spam, Spam, Spam, Spam...glorious Spam." The one called Mickey sang the old Monty Python tune and beat out a rhythm on the door with a badly damaged baseball bat.

"Shut up!" Yelled Dawg, followed by rapid fire expletives. "I can't hear myself think." Mickey swung the bat and its arced path ended on Dawg's skull with a crack. Mickey tossed the splintered and bloody bat to the floor taking command of the group. "I'm the Alpha now boys, and I say we eat! Like they say, 'It's a Mick eat Dawg world'... ha ha ha ha ha!"

That night, as the satiated bunch sat discussing how they would smoke out the Hardwicks using the air vents, three shots rang out from within the safe room. A wailful, mournful, cry was heard emanating from the vents. A fourth shot was heard a minute later followed by smoke and then flame.

San Martin residents didn't fair any better. While Sheriff Beecher, Deputy Don, and their appointed deputies held some semblance of order, there was no peace.

Clete Beecher and the citizens of San Martin had cooperated with HS, FEMA and the Army National Guard until they were called away one night to help out in Austin. Austinites had stormed the capitol building and the Governor's Mansion, starving, angry, and outraged at how they were being treated. Most of downtown Austin was engulfed in flames. The red-orange glow could be seen all the way in San Martin 25 miles away, and even at the space station. Thick, black smoke hung over the city choking off light from the moon and stars.

When real hunger settles in, ravishing, death rattling hunger, the mind is twisted. Coupled with end of the world scenarios, and world wide chaos, we have a prescription for stark raving madness.

The Sheriff and his deputies could only stall the inevitable. Like cities all across America and the world, San Martin's community broke down. Gradually, at first, and then completely.

The Hardwicks were home when marshal law was declared and before the military arrived, surrounding San Martin. They were never seen again by the good folks of San Martin.

James Jameson, reporter, was arrested for being an enemy combatant and inciting civil disturbance, then hauled off to the makeshift prison (later shot and dumped in the landfill).

Clete Beecher And Don Haynes tried to keep the townspeople unified, but in the end, were forced to either kill them or let them leave. Under the cover of darkness, people would go missing, and the next day the smell of meat, like sweet pork filled the morning air. Beecher and Haynes died of starvation.

As fate would have it, the professors, who predicted the end, were passed out drunk and never saw their own end coming. They never knew they were spam.

Harold 'Bugs' Smithton hid at Bug World, eating insects, until someone burned it down with him in it.

Around the world, people who were not killed and eaten, died of starvation and disease from the concentration camp-like city conditions. Then, once they were left to fend for themselves, from exposure in the outlying areas.

When the currency is food, even presidents, governors, and kings are on the same footing as everyone else. But, because they had more, they were the first to be sought out by the starving multitudes.

After cities were abandoned for the wildernesses, everything that moved was killed and eaten. Any plant life that could be eaten was pulled out of the ground, off of bushes and trees. And, yes, even old leather shoes were boiled and formed the basis of "hobo stew."

Humanity held on for a while, with isolated pockets of people here and there. Small fishing communities caught enough fish to survive until discovered by hordes, and mobs, and ravenous rabble. A few floating cities avoided the spammers, but were unable to overcome the mechanisms of extinction.

These thinned out until only single or pairs of individuals existed subsisting on whatever they could scrape up. Mankind was unable to recover. It happened so fast there wasn't any time for wars to reduce the populations.

All 104 U.S. nuclear power plants experienced "station blackout" when they lost access to off-grid electricity. Reactor cores were unable to be kept cool and core meltdowns occurred.

There was enough diesel fuel to power the backup emergency generators for seven days. Some plants had enough for about a month, as automatic shutdown of nuclear plants began.

The control rods dropped into the core, and water was pumped into the reactors. However, after the fuel ran out and plants were abandoned, water boiled off leaving the fuel rods exposed. Spent fuel pools of depleted radioactive materials didn't have time to decay and high density storage racks vented directly into the atmosphere as fires broke out. Multiply the scenario 435 times in the 30 countries world wide with nuclear plants, and one paints a very ugly picture.

Everything happened so fast that people were too busy trying to find food to think about a good old fashioned population reducing war. Only a couple of nuclear bombs, a few dirty bombs, and a handful of chemical bombs were set off. This was mostly by small governments and "terrorists."

Washington, along with the entire Capitol complex, disappeared beneath a mushroom shaped cloud.

When persons in undiscovered caves and underground bunkers came up to look for food, there was little left of the world they remembered.

Because of lack of genetic diversity, dwindling numbers of humans, disease and starvation, humanity was unable to recover and every human being on earth eventually succumbed to the mass extinction.

...

He wondered how the last human alive would feel. What would his or her last thought be? Dmitri watched from 205 miles above the earth, courtesy of the International Space Station. He saw a few trails of smoke beneath green tinted clouds on the light side of the earth, the glow of burning cities and an occasional flash of light as he passed over the dark side. Dmitri, himself, had passed over to the dark side days before. He grinned, with blood dripping down his chin, as he pulled the lever which opened the air lock.

Except for a few mammals in the oceans, mammals were history. Animals that walked upright no longer roamed the earth. Insects dominated the land and air. Fish dominated the seas.

The big fish ate the little fish, and an occasional insect that fell onto it's surface or sunk below. Once in a while fish floated to the surface and were washed ashore where they were consumed by carnivorous insects. But, for the most part, it was like there were two worlds. Fish tend to stay off of the land, and there are no insects in the oceans.

Although man was no longer around to pollute the land, the sea, and the air; to hunt, to trap, and to fish unabated, life in the sea faired no better. Pacific Sea Stars and Atlantic Starfish never recovered. Barnacles were better suited for the temperature and acidity of the water. The bacteria and viral pathogens didn't affect them as easily.

Soft bodied denizens of the deep were easily affected. Porpoise, dolphins, and whales succumbed to disease. There would no longer be any finger pointing and waving of hands. No accusations that man was the driving force of extinctions now. It was man who caused his own extinction. His intelligence and his technology being the driving force. Man was simply too efficient!

Animals and insects never blamed man. Fish eat fish, and insects eat insects. It's the way of the world. Man had to eat to live, and eat he did... until nearly every visible living thing that swam, paddled, or jetted; walked, scurried, scampered, crawled, burrowed or slithered; flew, jumped, ambled, or gaited...was gone. Every eatable tuber, fruit, legume or grass; every eatable bark, root and leaf... was gone.

But, unlike man, who would never rise again, seed, egg, larvae, and rhizome held the promise of a future for plants and insects.

Natural carbon dioxide pumps continued to raise the co2 levels until they were 1500 parts per million once again. It had been millions of years since such perfect conditions existed. Perfect for plant life! Earth became a lush green planet once again. Every flower, shrub, vine, bush, or tree proliferated.

The co2 released from the oceans and meeting the air at the surface, acidified the waters further. Rivers and lakes clogged with algae and hydra. The oceans filled with the Red Tide and other algae choking off light and oxygen, eating up all the nutrients and killing submerged aquatic vegetation, bleaching coral reefs, and suffocating the fish.

Warm waters, lakes, rivers, and streams filled with hydra and algae which entered into a symbiotic relationship. The multi-cellular hydra protected its single celled symbiote from predators, and the algae's photosynthetic products provided food for the hydra. Elsewhere, brown, cold water diatom algae multiplied rapidly.

Eventually, earth's waters were devoid of all life but algae, bacteria, and viruses. Fresh water from rain was quickly absorbed by thirsty plants and insects as it pooled on leaves and collected in depressions or divots in soil or rock. Insects entered into even greater symbiotic relationships with plants than before the fall of man.

The hydrocarbons in asphalt began to break down until streets were not even a memory. Iron and steel oxidized and cement crumbled, as it was covered in plant growth until the earth completely reclaimed the territory man had borrowed from her.

Wasps, bees, termites, beetles, and ants grew rapidly in numbers as they spread out into new territories. Warmer, moister climate was very favorable for the American Cockroach. Cockroaches left their caves and sewers to seek new homes above ground in such huge numbers, that their hissing could be heard for hours before they arrived in New New York or

New Chicago and New LA. The radiation had not killed them, but had actually made them stronger, and larger

The slower cell cycles, and fewer genes (and since they molted only once a week), made cockroaches and many other insects less vulnerable to radiation poisoning than humans and other mammals had been.

Beetles and fruit flies were even less sensitive, and they increased in numbers. But one of the life forms least effected of all, was previously known to humans as Conan the Bacterium. This really foul smelling, red bacterium, had originally evolved in cans of meat which had been radiated to preserve it. The radiation only doubled its lifespan.

Unencumbered by man, ants continued to grow in numbers until, pound for pound, they outweighed all other creatures combined. Slavemaker ants, Fire ants, Weaver ants, Carpenter ants, Leaf cutter ants, Harvester ants, Rasberry ants, Army ants, Wood ants; red ants, black ants, brown ants, yellow ants; predators, farmers, scavengers…

Where once there were 35,000 kinds of ants, a million ants for every man on earth, now, their numbers were uncountable, with new species exploding onto the scene. And they had voracious appetites. Although, the harvesters and leaf cutters were content with raising and eating fungus or stored grains, and other ants had fruits and vegetables, many others still had a hunger for protein. With the loss of mammals, they only had other insects and each other for food.

Although beetles made up more species than any other type of insect or bug, cockroaches were the greatest in number, second only to ants and termites. They had also increased in body mass considerably.

The oldest insect of all, dating back 350 million years before man, the cockroach, had outlived the dinosaur 150 million years and now had outlived man, but would he outlive the ant?

Rope Hypothesis
and
Thread Theory

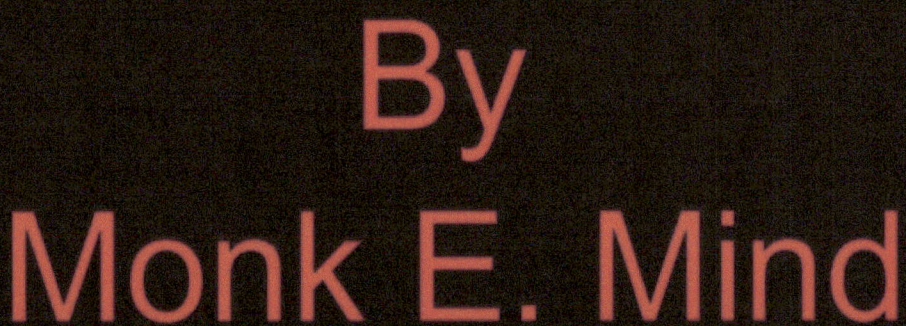

By
Monk E. Mind

www.ingramcontent.com/pod-product-compliance
Lightning Source LLC
Chambersburg PA
CBHW040410220526
45473CB00004B/1190